AF782158

Energy Marketing Handbook

ENERGY MARKETING HANDBOOK

by

DENISE WARKENTIN

PennWell Books

PennWell Publishing Company

Tulsa, Oklahoma

Copyright © 1996 by
PennWell Publishing Company
1421 South Sheridan/P.O. Box 1260
Tulsa, Oklahoma 74101

Warkentin, Denise
Energy marketing handbook/Denise Warkentin
p. cm.
Includes bibliographical references and index.
ISBN 0-87814-604-0
1. Electrical utilities—United States. 2. Electric power—United States—Marketing. 3. Natural gas—United States—Marketing. 4. Interconnected electric utility systems—United States. 5. Electricity wheeling—United States. 1. Title.
HD9685.U5W37 1996
621.042'068'8—dc20
Library of Congress
Cataloging-in-Publication Data 96-32407
CIP

All rights reserved. No part of this book may be reproduced, stored in a retrieval system, or transcribed in any form or by any means, electronic or mechanical, including photocopying and recording, without the prior written permission of the publisher.

Printed in the United States of America
1 2 3 4 5 00 99 98 97 96

DEDICATION

To my husband and life partner,

Keith,

who unconditionally believes in me
and supports me in everything I do.

CONTENTS

ACKNOWLEDGMENTS

This book was researched, written, and produced on a fast-track. I wish to thank several individuals and companies for enthusiastically supporting me and helping make this book project a little less taxing.

A special thank you goes to Jim Ferrier, PennWell Publishing Company's Information Products Group Acquisition Editor, for all his assistance both before and during the writing of this book. His advice and guidance was priceless.

Much gratitude goes to the following: Elaine A. Pare of The Arrington Group; Charles Walker Jr.; H. Joseph Fleischman, P.E.; Robert Driscoll; Michael D. Carlisle; National Independent Energy Producers; Robert Smock; Wayne Beaty; Nachamah Jacobovits of New York Mercantile Exchange; Boston Edison; Lindsay von Ulm Paulson of Cegelec ESCA; Peter C. Christensen; UtiliCorp United/EnergyOne; Calpine Corporation; Parks Associates; National Regulatory Research Institute; R.J. Rudden Associates Inc.; and Washington International Energy Group.

FIGURES

TABLES

THE ENERGY MARKETING HANDBOOK

FOREWORD

Making and marketing electric energy once was a rather simple business. Electric utilities, in times past, had their prescribed service territories with little or no competition. Marketing plans meant trying to attract businesses and industries to locate in a particular service territory.

Deregulation and opening up the power delivery systems to non-utility generators has ushered in a new era of fierce competition among electricity suppliers and electricity users—especially the large industrial and commercial users.

A new class of players—energy marketers and energy brokers—has entered the competitive arena and are ready to wheel and deal.

Denise Warkentin has captured the essence of the era of energy marketing in this new, much-needed book. She expertly lays the historical foundation and then in a logical and understandable fashion leads us into the intricacies of how energy marketing is reshaping the industry and who will most be affected and how.

Warkentin has an excellent grasp of how the industry has functioned in the past as a regulated industry and how the changes are now reshaping this once staid industry into a dynamic, fast-changing business that will be dependent on good marketing techniques.

The Energy Marketing Handbook is a must for anyone involved or interested in the electric utility industry.

Wayne Beaty
Managing Editor
Electric Light & Power

PREFACE

The essence of *change* brings about endless, untold opportunities. That said, most people reading this book who are involved in some way in the energy business probably are thinking about what exactly a book on energy marketing would accomplish. The main mission is that of defining, explaining, and expounding on some critical aspects associated directly and indirectly with the marketing of energy as a commodity.

The marketing of energy represents such a *change* in the thinking and in the business activities of energy providers. Opportunity abounds when it comes to looking into the crystal ball to at least try to foretell—even speculate—what the future beyond holds for the competitive energy business.

The not-in-my-back-yard philosophy that seemed at one time to go hand-in-hand with the electric utility industry (and also with the now-deregulated gas industry) surrounding the notion that it *may* some day be among the ranks of the deregulated, competitive industries, has been buried—deep.

TABLE PRE.-1 1996 MARKETPLACE CHANGES

Question 1: Please identify the marketplace changes your organization is facing. Rank each based upon its anticipated impact on your organization.

Marketplace Change	Very Significant		Significant Impact		Little or No Impact		
	Number Responded	% of Respondents	Number Responded	Percent of Respondents	Number Responded	Percent of Respondents	Total
Market Segmentation	70	42%	78	47%	17	10%	165
Industry Consolidation	63	38%	67	41%	35	21%	165
Energy Conservation	6	4%	48	29%	111	67%	165
Beyond-the-Meter Products/Services	50	30%	87	53%	28	17%	165
Pricing Pressure/Non-Tariffed Pricing	94	57%	56	34%	15	9%	165
Alternative Energy Sources	16	10%	61	37%	88	53%	165
Cogeneration, Self Generation	28	17%	80	48%	57	35%	165
Independent Generation	31	19%	78	47%	56	34%	165
Power Brokering	70	42%	71	43%	24	15%	165
Emergence of Retail Energy Marketing Companies	66	40%	70	42%	29	18%	165
Retail Wheeling	85	52%	54	33%	26	16%	165
Wholesale Wheeling (tranco separation) (FERC)	71	43%	74	45%	20	12%	165
Other Regulatory Changes	42	25%	86	52%	37	22%	165

Source: Reprinted with permission by The Arrington Group, Copyright 1996.

Change has descended upon the energy industry. It is up to you and others involved in this business of energy marketing to see to it that the opportunity to make this space of what will one day become history, the best and most fruitful it can be.

To better illustrate the many changes that are upon the industry, Table Pre.-1 and Table Pre.-2 list some of the most significant changes that are occurring in the industry.

The Arrington Group, a management education and consulting firm in the field of organizational transformation, obtained the results shown in both of the tables by polling senior-level management at major investor-owned electric utilities in the United States.

Table Pre.-1 contains 1996 data, while Table Pre.-2 contains 1995 data. The two different tables are presented here to illustrate how just one year can alter perceptions about which marketplace changes are significant and which ones are seen as the most significant.

TABLE PRE.-2 1995 MARKETPLACE CHANGES

Question 1: Many marketplace changes are apparent. Plaese identify the 5 most significant changes from your perspective.

ITEM	# of Respondents	RANK #1	#2	#3	#4	#5
Increased Competition	50	34	5	2	3	6
	98.04	*66.67*	*9.80*	*3.92*	*5.88*	*11.76*
Increased Customer Pressures	26	6	8	3	4	5
	50.98	*11.76*	*15.69*	*5.88*	*7.84*	*9.80*
Conservation	2	0	0	1	1	0
	3.92	*0.00*	*0.00*	*1.96*	*1.96*	*0.00*
Recuced Demand	5	1	2	0	1	1
	9.80	*1.96*	*3.92*	*0.00*	*1.96*	*1.96*
Cogeneration	8	1	2	2	0	3
	15.69	*1.96*	*3.92*	*3.92*	*0.00*	*5.88*
Beyond The Meter Service	9	0	2	0	2	5
	17.65	*0.00*	*3.92*	*0.00*	*3.92*	*9.80*
Outsourcing	6	0	0	2	1	3
	11.76	*0.00*	*0.00*	*3.92*	*1.96*	*5.88*
Productivity	14	2	4	3	3	2
	27.45	*3.92*	*7.84*	*5.88*	*5.88*	*3.92*
Pricing Flexibility	24	2	4	8	7	3
	47.06	*3.92*	*7.84*	*15.69*	*13.73*	*5.88*
Regulatory Changes	27	5	4	8	8	2
	52.94	*9.80*	*7.84*	*15.69*	*15.69*	*3.92*
Wholesale Wheeling	26	4	6	7	5	4
	50.98	7.84	11.76	13.73	9.80	7.84
Retail Wheeling	33	4	11	7	8	3
	64.71	*7.84*	*21.57*	*13.73*	*15.69*	*5.88*
Reduced Connect Time	5	0	1	1	2	1
	9.80	*0.00*	*1.96*	*1.96*	*3.92*	*1.96*
Independent Generation	23	4	3	7	6	3
	45.10	*7.84*	*5.88*	*13.73*	*11.76*	*5.88*
Power Futures Market	4	0	1	2	0	1
	7.84	*0.00*	*1.96*	*3.92*	*0.00*	*1.96*
Fuel Cells	3	0	0	0	2	1
	5.88	*0.00*	*0.00*	*0.00*	*3.92*	*1.96*
Other:	5	1	0	0	1	3
	9.80	*1.96*	*0.00*	*0.00*	*1.96*	*5.88*

Corporate transformation
Energy brokering/marketing
Power brokers & marketers
Power marketing
Streaming plants

Source: Reprinted with permission by The Arrington Group, Copyright 1995.

Peruse through both the 1996 and 1995 lists of most significant changes as reported by respondents to the Arrington Group study. Notice the concerns that are termed *significant*—at least by today's standards—are vastly different than the concerns of the past. While those concerns had their place, so too do the new ones. Suffice it to say that increased competition was *not* among the concerns of the past.

Things are changing, evolving. All concerns, whether they are those presented in the past or those that are being presented currently and on into the future, all have merit. We would all do well to heed these new concerns and prepare now for the industry's new operating agenda.

ENERGY AS A SERVICE?

Energy is no longer thought of as a service. Getting used to this idea is about as hard as getting used to the resurgence of 1960s apparel.

Whether you're talking about varieties of energy that encompass electric, natural gas, cogeneration, or others, the delivery of energy as I will refer to these three forms of energy in this book, is better known in the current age as the delivery of a product. And, because most all products keep businesses and people who live in the civilized and privileged world happy and productive, the product of energy fits the bill quite nicely.

Products cost somebody, somewhere. And, most assuredly, somebody, somewhere will pay the cost. If that person didn't pay the cost then that somebody, somewhere would not have that product, or in some cases, would choose another product—perhaps with a better price or some other more attractive quality.

As in the case of energy, there is little choice but for somebody, somewhere to pay the cost. We all use energy every day in our lives—in our homes, our vehicles, our offices, etc.

The attractive thing about products—which after all, are essentially raw products that are used, for example in the case of energy, to produce electricity in the light switch or lamp you turned on to read this book—is that their cost can be foretold.

ENERGY AS A COMMODITY?

Much like oil and natural gas, electricity is a commodity. As a commodity, it should come as little surprise that electricity—just like oil, natural gas, and other forms of energy, which are all considered *commodities*—can be sold in the cash or spot market.

Amazing, isn't it? It's hard to imagine actually being able to buy and sell energy much like is done currently with petroleum, natural gas, and oil. Even the bran in your cereal and the wheat in your bread are bought and sold on the open market (and in the electricity trading pit, no doubt) on the floor of a major exchange, like the New York Mercantile Exchange (NYMEX).

NYMEX launched its Palo Verde and California-Oregon Border Electricity Futures Contracts on March 29, 1996. NYMEX began offering the contracts for electronic trading on NYMEXACCESS^{SM} on April 1, 1996 as part of its April 2, 1996 trading session.

NYMEX President R. Patrick Thompson said, "We have not seen this level of interest in a new product since we launched natural gas futures, which went on to become the fastest growing contract in Exchange history."

Being such a profound and unprecedented event in the history of futures contracts, the NYMEX electricity contracts surely cut to the chase about what the future holds for the marketing of energy as a commodity.

THE JOURNEY ONWARD

In a first-ever attempt to get all my thoughts—in one place—about the subject of energy marketing, I have, through the pages of this book, attempted to undertake this feat in the most non-technical way possible.

You will learn through reading this book, in addition to what the term *energy marketing* really means, background about why competition is happening in the electric utility industry, how you can serve as an enabler of change in this vastly evolving industry, and how you can reduce—and even manage—your risk when buying and selling electricity on the open markets. Through reading this book, you also will learn the processes that are involved in buying and selling energy using new technological advances, and how and why futures contracts can be used to optimize market power and presence.

In addition, because so much of what we learn as individuals and as business people is learned through the experiences of others, chapter 3 has been devoted to the unique interplay that now exists between the electric utility industry and the natural gas industry.

Sit back, relax, and enjoy your journey through the amazing and sometimes mind-boggling world of the marketing of energy. I hope you enjoy reading this book as much as I enjoyed writing and presenting it to you.

Here's to the close-at-hand future—the future of energy marketing.

CHAPTER 1

INTRODUCTION TO ENERGY MARKETING

Energy, as referred to in this book, encompasses the broad spectrum of energy supply in the United States and abroad. Energy not only includes electric power but also natural gas, cogeneration, self generation and a host of other energy alternatives available in the market today.

Due to space limitations and the complexities involved in covering every energy source adequately, a great majority of this book is devoted to electricity. This emphasis on electricity should in no way be construed to mean that other forms of energy are viewed as any less important or integral to the overall process involved in the marketing of energy.

MARKET DETERMINES ENERGY SOURCE

The phrase "energy marketing," by its very nature, involves the rationalization and evaluation of all energy sources by various parties (the electric utility, natural gas utility, independent power producer, cogenerator, and end-user, etc.) to determine the best overall fit in attending to customer needs.

Energy encompasses the broad spectrum of energy supply in the United States and abroad.

The commodity itself—the energy—is basically becoming a secondary concern to most stakeholders as needs themselves become primary in the choice of which type or types of energy will be utilized. The market itself, in many instances, will help to determine the source from which energy is produced.

A CUSTOMER-DRIVEN MARKETPLACE

Energy customers, once thought of as a captive audience due to the monopoly position of their energy supplier, can no longer be held captive. Customers want choices when it comes to which energy sources they utilize and are provided. Customers are demanding that they be given the right to make these choices—now. The accompanying sidebar, "What Utility Executives Fear Most: Loss of Industrial Customers," provides testament to the fact that utilities already are concerned about keeping their existing industrial customers happy with the services and the prices they are able to provide.

What Utility Executives Fear Most:
Loss of Industrial Customers

Electric power executives are more concerned than ever about losing valuable industrial customers. The more pessimistic outlook likely reflects increasing pressures from key customers. Although 69% of respondents felt that *public* pressure for reducing rates will increase, 90% felt that *industrial* pressure would. There is a clear expectation that such pressure will result in greater discounting for large industrials (80% agree) and more differentiation of prices among customers within the same traditional classes (72% agree).

Research among industrial customers confirms that utilities' fears are well founded. In fact, if the expectations of many large consumers come true, actual load losses may be more severe than feared. As the rhetoric of deregulation grows, so do consumer expectations. Large industrial users are increasingly optimistic about their ability to win cheap electricity rates. At the same time, they are impatient with the pace of deregulation. Industrials see themselves as commodity buyers who should enjoy the benefits of competitive procurement in a robust market. As large consumers avoid long-term relationships in favor of flexibility, the stability and reliability offered by utilities will no longer be of significant value.

Customer feedback also suggests that utilities may be burdened by the industry's past relationship with consumers and by its traditional image. Many large customers hold a negative perception of the industry and tend to characterize utilities as inflexible and inefficient, as well as unable and unwilling to cope with growing competitive forces. Although they acknowledge the positive intentions of some utilities, they nevertheless believe that most utilities, even the most aggressive, have yet to demonstrate an appropriate customer orientation.

A competitive advantage will accrue to those utilities that distance themselves from the traditional image and demonstrate a more progressive, customer-focused identity.

Industrial & Large Commercial Customers

What They Want:	What They Do Not Want:
Lower Prices	Formal High Rates
Right to Choose	Lack of Choice
Easy Reliability	Sticky Relations with Utilities
Less to Worry About through Outsourcing & Risk Sharing	
Single Shared Billing	
Company-Wide Pricing	

Source: Reprinted with permission by the Washington International Energy Group.

Not only are energy customers desiring the opportunity to make choices about their providers, but also they are requiring (and will require) their supplier or suppliers to assist them in addressing their energy *needs.*

This currently can be said to be true of large industrial and commercial consumers of energy. However sometime in the future—whether near or far, who knows?—residential consumers also may be seeking help in addressing their energy needs.

This spells out the need for energy marketers to be solution providers and not just merely a *provider* and also not just a provider that offers the best prices.

A MARKETING MODEL

In early 1996, Edison International (formerly SCEcorp.) launched Edison Source, an unregulated energy marketing subsidiary. And although rather new it has burst onto the scene with some model goals in mind.

Touting itself as more than just a power marketer, Edison Source will focus on two main customer needs and concerns: energy efficiency and energy infrastructure improvements and electrotechnologies that help solve air quality, water treatment, and other environmental challenges.

According to the company, Edison Source will work to blend energy solutions in order to help consumers understand the big picture where all aspects of energy use are used to the strategic advantage of a company's buying power.

From purchase and transportation of the energy to capturing the greatest efficiencies of energy in manufacturing processes, Edison Source has vowed to help its consumers trouble-shoot so that consumers get the best value possible.

Value really is the bottom line in energy marketing. Energy companies should be asking themselves, "What value or values can I offer my customers" and "What activities do I need to engage in to get these values transferred to my customers?"

I think Edison Source is just one of several excellent subsidiaries that have been formed by countless numbers of utilities (both gas and electric) to help address these burning questions.

There are many challenges and opportunities that exist in the world of energy marketing. The following chapters attempt to address the plethora of ingredients that have come together to make this activity the talk of the decade.

CHAPTER 2

Evolution In The Energy Industry

Like most all other industries, the energy industry in its current state is the product of evolution—a gradual, yet continual state of advancement brought about through various events.

The evolution that has occurred in the electric utility industry is also correctly characterized as having gone through a progressive historical *transformation*—from an unregulated industry (from around 1879 to the 1900s) to one of regulation that began in mid-1930s and persists into the present time.

However, current efforts aimed at deregulating the electric utility industry suggest that the regulation that has persisted for almost a century is nearing its final destination. These deregulation efforts, which are discussed in detail later in this chapter, are unprecedented and are due to have various repercussions on the power industry and on other industries as well.

THE EFFECT OF HISTORICAL EVENTS

In the context of evolution, it can be said that historical events have brought pressure to bear on the electric power industry to make significant, profound changes in the way the industry conducts its business. By any measure, these historical events have been a great catalyst in the movement of the power industry toward competition. These events are detailed later in this chapter.

Evolution by its very nature, context, and definition, is ongoing. It never ends. It is constant. Indeed there may be times when things come to a standstill and present the overall impression that change or the process of change (which is at the heart of evolution) has ceased.

It also may *seem* that changes in, for example, the electric power industry, occur rather quickly—that the changes just appear as if by magic. But, they don't.

A series of historical events has actually *caused* the industry to *evolve* and to change into something else. Outside forces, including regulatory policy and, in no lesser part, the economy are direct agents in causing the power industry to change. The same can be said for the natural gas industry.

DEREGULATION AND THE ROLE OF EVOLUTION

Deregulation, in its truest form, is the process of refashioning or retooling existing rules and regulations.

It's easy to think of examples where age-old industries have been retooled into so-called *new-age* industries, reflecting the trends of the current and future time. What comes of it is termed *deregulation.* Examples of deregulated industries include, besides natural gas, railroad, banking, telecommunications, and airlines. The electric utility industry is the last major U.S. industry to be deregulated.

Some Controls will Remain

One thread that seems to bind all the major deregulated industries, which includes electricity because it is well on its way to deregulation, is that *some* form of regulation tends to remain behind.

Take, for instance, the natural gas industry. Most wholesaling functions were deregulated during its restructuring. Other aspects of gas industry regulation were largely left untouched by formal deregulation. Interestingly enough is the fact that the Federal Energy Regulatory Commission (FERC) has begun deregulating different aspects of the gas industry by special rules. These other aspects FERC is working to loosen from regulatory bondage thus far are not supported by any legislative authority.

In the electric power industry, deregulation will more significantly affect the transmission side of utility businesses than that of generation. This is due to the fact that FERC, for the time being, is deferring to the direction Congress set out as it pertains to generation in the Energy Policy Act of 1992. Generation, according to this act which is discussed later in this chapter, has few regulatory controls.

Another area where regulatory control will be felt even though deregulation will occur (and is occurring) is at the distribution end of the electric business. Distribution largely falls in the regulatory purview of state regulatory commissions.

Starry, Starry Night

To *totally* deregulate any industry—including the electric power industry—would be to do away with *all* controls and watchdog functions of the federal government, and that would be most assuredly not in the stars, as they say.

In fact, it can be said that evolution simply does not allow for a so-called copycat version of what *was*. Due to the evolution of these two industries and taking into account the gradual and deliberate events that have taken place over the years, it seems virtually inconceivable to think that a true form of deregulation could ever occur.

All-out deregulation is inconsistent with cause and effect. It is inconsistent with the whole theory of evolution—that things change because outside forces (namely events) have caused them to do so. You cannot go back to square one where absolutely *no* regulation existed.

Because an examination of the historical nature of both the gas and electric power industries is in order to fully understand how these industries and their markets (and thus the energy market) are truly the subject of the evolutionary process, the following presents a somewhat abbreviated version of the colorful advancement of each industry.

THE EVOLUTION OF THE NATURAL GAS INDUSTRY

Natural gas regulation began in 1938, with the passage of the Natural Gas Act. However, before that time, the natural gas industry enjoyed relative freedom from government controls.

Partial deregulation was set into motion in 1985 with FERC's passage of two landmark orders. FERC Order 436 was passed to help address and control the apparent discriminatory access to pipeline transportation. The courts later would not completely disagree with the purpose of the order, but would, however, have some differences of opinion on how to address the discriminatory access dilemma.

The second major step was the passage of FERC Order 451, which effectively removed vintage price tiers that had been set out under the Natural Gas Policy Act (1978). With this order, FERC raised ceilings of vintage gas categories to market levels and established procedures that encouraged producers and pipelines to renegotiate associated contracts.

Due to some legal challenges that faced FERC Order 436, it was reentered into the deregulation scheme as FERC Order 500. This rule would later be finalized and be renamed FERC Order 636, which became the gas industry's ticket to deregulation.

As will be detailed later in this chapter, the FERC's issuance of its final rule on open access for the electric power industry on April 24, 1996—just four years after the gas industry was deregulated—started the deregulation wheels to turn for electricity.

THE EVOLUTION OF THE ELECTRIC UTILITY INDUSTRY

The Public Utility Holding Company Act or PUHCA came into being in 1935 when Congress decided that it was time to restrain the activities and dealings of large holding companies in both the natural gas industry and in the electric power industry.

There is a question over whether PUHCA,
should die a quiet death to mark
the beginning of a new era in the electric power industry.

Due to a number of changes in the electric utility industry, the purpose of PUHCA has come under fire by industry stakeholders—including legislators, regulators, and others—questioning whether PUHCA is still needed under current, and expected future, conditions.

There is a question over whether PUHCA, which has lived its life during the last several decades as a means to protect investors, consumers, and the general public, should die a quiet death to mark the beginning of a new era in the electric power industry. As competition in the industry looms, there are some provisions in the aged structure of the act that some utilities and others feel actually will hinder the progression of a market-driven industry.

THE HISTORY OF PUHCA

Large monopolistic empires that were largely inadequately financed and secured began cropping up during the early 1900s when interstate holding companies were acquiring electric and gas utilities at inflated prices.

The 1929 stock market crash was essentially the point of light for many of these holding company abuses as many investors lost millions of dollars. Many holding companies went into bankruptcy.

More than 67% of the electric generation in the United States was controlled by eight major holding companies by 1932. Congress began mulling over what it should do to halt the apparent abuses. In response, it passed PUHCA.

The goal of PUHCA was to simplify and regulate the organization of holding companies and to limit the geographic size of utilities. The act gives the Securities and Exchange Commission (SEC) authority over those who hold interests in utilities and restricts the operations of *registered* public utility holding companies.

PUHCA introduced limitations on the activities, and the actual structure, of companies that directly or indirectly own or control 10% or more of a public utility. The only holding companies exempt from the strictures of PUHCA are those that have limited their scope to a single state or can meet other such criteria.

An amazing fact is that while there are upwards of 100 utility holding companies, only about 15 electric and gas utilities are *registered,* which means that they are fully regulated under PUHCA.

Those companies that are considered exempt from PUHCA are still subject to state regulation, but are not governed by special SEC requirements. The only SEC requirement that does apply to them, however, is one that states that even exempt holding companies cannot own utility company interests outside their geographic area or acquire more than 5% of a utility's securities without the approval of the SEC.

Exempt companies are able to get into other lines of business, if they so choose. Registered companies are prohibited from doing so. Registered companies are strictly confined to their single, integrated business, which may reach into other businesses, but those "other" businesses must be "reasonably incidental" or "economically necessary."

Further, registered holding companies cannot be involved in any way in utility or non-utility businesses, except under certain and limited conditions. They also cannot invest in utilities that are not a part of their integrated system.

Other such constraints on the few *registered* public utility holding companies that exist include provisions that they cannot do the following, unless SEC approval is sought:

- Participate in any contracts associated with service, sales, or construction when the contracts are with affiliated public utility companies.

- Issue or sell equity securities or alter the priorities, preferences, voting power, or other rights of any existing securities.

- Acquire securities, utility assets, or other interests in any other business.

- Borrow from subsidiary companies or receive credit from any subsidiary public utility company.

- Declare dividends.

Efforts to Repeal PUHCA

As can be seen from the discussion of PUHCA's constraints, many utilities have the opinion that the limitations imposed by PUHCA under a traditional, monopolistic industry structure, are acceptable. However, due to the vastly changing nature of the electric power industry, PUHCA has become unnaturally constraining.

It is generally thought that the restrictions imposed by PUHCA will actually cost registered public utility holding companies business via restricting their ability to make strategic decisions quickly.

However, on the other side of the coin, there are many in the industry who feel that PUHCA repeal (as well as repeal of the Public Utility Regulatory Policies Act, as discussed below) can only be repealed when the electricity industry reaches full competition. This is believed by some stakeholders because of the feeling that PUHCA would prevent some of the abuses that may occur as the industry moves toward deregulation.

In 1982, the SEC decided that the set-out purposes of PUHCA had been achieved and that it was time to seek its ultimate repeal. In recommending this action, the SEC also suggested that certain functions of the act be transferred to the authority of FERC. However, in 1995 the SEC again brought up the issue of repealing PUHCA in the form of a 150-page review of the act. Congress has yet to act on the SEC's recommendations.

A bright day was had, however, when, in 1992, the Energy Policy Act (EPAct) was passed by Congress and some of the tighter restrictions on holding companies were loosened. A more detailed discussion of EPAct is presented later in this chapter.

PUBLIC UTILITY REGULATORY POLICIES ACT

Enacted by Congress in 1978, the Public Utility Regulatory Policies Act or PURPA encourages the use of renewables to generate electricity. The goal of the act, passed at the height of an energy crisis, was to help the United States decrease its dependence on foreign oil. It still exists, and to many in the industry is a dated act that has caused severe problems, especially in the area of non-utility power projects.

A section of the act, Section 210, encourages private investors to build small electric power production facilities using renewable energy sources (such as sun, wind, and solid waste, etc.). Also, PURPA encourages plants to cogenerate electricity by using excess steam from various manufacturing processes.

These private investors who built the small plants using renewables were termed "qualifying facilities" or QFs. And, if QFs would develop PURPA projects, a market was guaranteed because the act mandated that utilities in need of generating capacity buy all the electricity QFs had to offer.

Most utilities have discovered that they are, in fact, having to pay more for, or pay for unneeded power, from QFs.

The Twist Behind the Regulatory Punch

Until now, the act seems perfectly logical. However, there is a twist. PURPA states that utilities will pay no more for generating capacity purchased from QFs than they (the utilities) would pay if they had generated the power themselves—otherwise known as a utility's "avoided cost."

In actuality, most utilities have discovered that they are, in fact, having to pay more for, or pay for unneeded power from QFs. Whether right or wrong, the opponent voices are being heard across the country: "Repeal PURPA to ensure that a competitive electric generation market can exist and customers can get the lowest price possible in a competitive market."

PURPA Constraints

Many in the industry are pushing for the repeal of PURPA. However, as pointed out earlier, some stakeholders believe that PURPA has its place to prevent abuses as the industry makes its way toward deregulation.

Still, many in the industry feel the act no longer makes sense and should be taken off the regulatory books. A competitive environment, repeal proponents say, will not support the act and its intent. In 1995, Sen. Don Nickles, R-Oklahoma introduced a bill, S. 708, which seeks to do away with most of PURPA's provisions.

The main complaints against the act are that PURPA projects:

- Cause less flexibility in planning;
- Hinder utilities' ability to purchase electricity at the least cost available;
- Actually discourage the use of renewable energy sources;
- Do not provide equitable rates to consumers; and
- Threaten reliability.

THE NATIONAL ENERGY POLICY ACT: THE TURNING POINT

Seen as a turning point by many in the evolution of the electric power industry was Congress' passage of the National Energy Policy Act of 1992 or EPAct.

The passage of EPAct would set the stage for competition in the electric utility industry.

Significant alterations were made to PUHCA with the passage of EPAct. The most important change was aimed at loosening the tight grip PUHCA had on registered holding companies. EPAct would begin allowing registered public utility holding companies to diversify into such areas as independent power production as well as others.

The passage of EPAct would set the stage for competition in the electric utility industry. Its thrust was to promote economic development and energy conservation, but it had a major goal of *encouraging* competition. EPAct began what has become a tidal wave of changes in national policies and regulations that have, in the past, defined the place of utilities in the generation, transmission, and sale of electricity.

New Class of Power Producer Created

A significant step taken by this "landmark" act was that it revised PUHCA to include a new class of power producer, called exempt wholesale generators or EWGs. This new class of producer, the act states, may be owned by a utility, a utility holding company, or by a developer that is not necessarily connected to a utility.

The generation market, under EPAct, is significantly freed up. EWGs are able to compete with experienced utilities that have been in the electric generation business for years. What's more, the act permits *registered* electric utility holding companies to operate an EWG anywhere in the United States and to reach out into foreign markets to purchase or develop generating facilities.

New Challenges Presented by EPAct

EPAct also has created some challenges for the electric utility industry. In the area of transmission, the act mandates that utilities allow outside wholesale electricity generators access to their power lines.

FERC has the power to order utilities to comply if FERC can discover that certain conditions have been met. The transmission of power for a third party is termed "wheeling" and will be addressed in chapters 4 and 5.

EPAct strictly prohibits, however, FERC from ordering "retail wheeling," which is the transmission of power to an individual customer for a generator of electricity other than the utility. FERC is also prohibited from ordering wheeling for a seller of electricity who does not have a firm contract from a party who wishes to purchase power.

FERC ISSUES HISTORIC FINAL RULES

On April 24, 1996, FERC issued a gargantuan set of final rules, known as FERC Order No. 888 and FERC Order No. 889, pertaining to electric utility industry deregulation. The FERC orders signaled to the rest of the world and to other industries that all-out regulation is not what it wishes upon the power industry. The wheels of change in the power industry are causing a once staid industry to begin a new adventure into becoming competitive *and* becoming viewed as a commodity-driven product.

The Open Access Final Rule: FERC Order 888

FERC Order 888, the result of many months of debate over a FERC electric utility restructuring mega-NOPR (notice of proposed rulemaking), opens wholesale power sales to competition. Public utilities that own, control, or operate transmission lines are required to file non-discriminatory open access tariffs that offer others the same transmission service they provide to themselves. This will:

- bring lower cost power to electric consumers;
- ensure continued reliability of the electric power industry;
- and, provide for open and fair electric transmission services by public utilities.

Pro Forma Tariff Provision

FERC, as part of Order 888, issued a single pro forma tariff, which describes the minimum terms and conditions of service to bring about non-discriminatory open access transmission service.

Under this provision, all public utilities that own, control, or operate interstate transmission facilities are required to offer service to others. The pro forma tariff also stipulates that utilities also must use the tariffs for their own wholesale energy sales and purchases.

Stranded Cost Recovery

To the delight (and relief) of public utilities, FERC Order 888 grants full recovery of stranded costs. FERC stated that stranded costs were those costs that were prudently incurred to serve power customers and that could go unrecovered if these customers use open access to move to another supplier.

As far as eligibility is concerned, FERC stated that recoverable stranded costs are those that are associated with wholesale requirements contracts signed before July 11, 1994. After this date, the rule stated, recovery must be provided for in contracts. Stranded costs will be recoverable from a utility's departing customers, under the new FERC rule.

Same-Time Information Systems: FERC Order 889

A second rule, Order 889, was also issued by FERC on April 24, 1996. This rule is known as the Open Access Same-time Information

System rule (OASIS). OASIS will ensure that transmission owners and their affiliates do not have an unfair competitive advantage in using transmission to sell power.

The OASIS rule requires public utilities to:

- Obtain information about their transmission system for their own wholesale power transactions in the same way their competitors do through an OASIS on the Internet; and,

- Separate their wholesale power marketing and transmission operation functions.

New Tariffs

A third item FERC took action on was the issuance of a NOPR on the Capacity Reservation Open Access Transmission Tariffs or CRT. Under the CRT NOPR, each utility would replace the Open Access Rule pro forma tariff with a capacity reservation tariff by Dec. 31, 1997. The proposal stipulates that utilities and all other power market participants would reserve firm rights to transfer power between designated receipt and delivery points.

FERC stated that its proposed reservation-based service seems to be more compatible with the open access requirements that market participants know how much transmission is available for electric power purchases and sales. FERC also stated that it also may be more accommodating to competitive changes going on within the industry.

PACESETTERS FOR THE REST OF INDUSTRY

As can be seen from the foregoing historical refresher of the electric power industry, it becomes clearer than ever that historical events undoubtedly have brought much pressure to bear on the industry to change. And, it already has.

One of the most compelling illustrations of how the industry is responding to its history and making changes occurred, and is occurring, is in the state of New Hampshire. This state has passed a law that requires full implementation of retail wheeling for all customer classes by Jan. 1, 1998.

In addition, the state of California not only has become a trend-setter with a statewide deregulation plan it has proposed, but also has become the "breaker of the ground" for a *national* plan.

The state's public utility commission (PUC) in 1995 adopted a plan, otherwise known as the "blue book" on electric utility industry restructuring in California, to create a competitive utility market structure. The plan provides for competition and consumer choice beginning by 1998 and would be the first statewide deregulation program in the nation. The PUC's vision, under this plan, is to have all customers actively participating in direct access by 2003.

The PUC has held numerous hearings and comment periods on the plan. A decision by the California Legislature is pending.

Then, shortly after FERC issued its final order on restructuring in the electric utility industry, several California electric utilities, the state's governor, consumer advocates, small business advocates, customer representatives, and public officials made suggestions to federal regulators on how to implement a new competitive electric market structure.

The listed stakeholders' vision is to form a power exchange (called WEPEX) and an independent system operator. Through the exchange, generators would be able to compete against each other to offer the lowest cost power to consumers. Chapter 7 details this timely and precedent-setting announcement further.

CHAPTER 3

The Interplay Between Gas And Electricity

Natural gas and electricity, while having some traits in common, are at base very different. Electricity, as well as natural gas, has its own set of highly unique characteristics, as shown in Table 3-1. However, their similarities do merit mention. Both have a growing focus on meeting customer needs; both industries are undergoing an increased number of mergers and acquisitions; and both have commodities markets.

Despite the unique nature of each industry, there still seems to be a kernel or two of truth in the statement that what the gas industry went through during its deregulation, while not exactly serving as a "model," can at least be testament to the electric power industry that deregulation can work. Perhaps, most importantly, the natural gas industry has taught that deregulation can lower energy prices and thus satisfy many end users' desires for more competitively priced energy.

NATURAL GAS AND ELECTRICITY AS COMMODITIES

Electricity and natural gas in terms of *exchange-traded commodities,* possess some very similar qualities.

To illustrate, consider that both natural gas and electricity, as do most all exchange-traded commodities, have a wide array of both buyers and sellers, are most generally exchangeable or replaceable for a similar type product or at least a product that could deliver the same end-use, and possess volatile price structures.

It can be said that natural gas has very high price volatility, and electricity is expected to have increasing price volatility as deregulation and competition in the industry progress.

THE GAS INDUSTRY VERSUS THE ELECTRIC UTILITY INDUSTRY

In May 1994, the National Regulatory Research Institute (NRRI) issued a study, "Overview of Issues Related to the Retail Wheeling of Electricity." In the report, the NRRI cautions against using the natural gas industry experience with deregulation as a *model* because of the fact that the electric power industry has some major variances in structure, practice, and policy.

As shown earlier in Table 3-1, "Comparison of Electric Power and Natural Gas Industries," NRRI came up with a list that shows very little

Table 3-1 COMPARISON OF ELECTRIC POWER AND NATURAL GAS INDUSTRIES

Electric Power Industry	Natural Gas Industry
1. Highly vertically integrated	1. Different owners of production, transmission and distribution assets
2. Nonstorability of electricity	2. Substantial gas storage capability
3. Need for backup electric service (unavailability of momentary substitute for electricity)	3. Fuel-switching capability by industrial customers
4. Multi-functional use of electric transmission lines	4. Unbundled characteristics of transportation service
5. Electric utility could lose substantial profits when customers buy electricity from another supplier	5. Gas distribution lose little or no profits when customers buy gas from another supplier
6. Loose control of electricity transmission in interconnected system	6. Precise control of gas transmission
7. Potential for large stranded investments with retail competition	7. Little stranded investments with retail competition

Source: Reprinted with permission by the National Regulatory Research Institute, Copyright 1994.

similarity—in structure anyway, and that's what we're talking about right now—between the two industries.

Structural Differences Exist

There are several reasons, which will appear in the next few paragraphs, that it has been difficult to structurally align the natural gas industry with the electric utility industry. As such, it can be argued that paralleling the deregulation of each of the industries (currently going on in the electric sector) is sometimes difficult.

Perhaps the biggest reason it has been difficult to align the two industries is that the electric utility industry has been vertically integrated—at least traditionally. Vertically integrated means that utilities sell power on a bundled basis, with utilities' activities running the gamut of the different functional activities of generation, transmission, and distribution. It must be noted, however, that this is changing. Chapters 1 and 2 provide the details on how and why vertically-integrated electric utility companies are on their way out.

The electricity industry is differentiated from the natural gas sector in that the gas industry already is unbundled. Production, transmission, and distribution assets are already owned by different entities.

Vertical Disintegration

A good example of how some electric utilities are already planning for future competition through vertical disintegration comes from Dallas, Texas-based Central & South West Corp.

The utility announced in mid-April 1996 that it soon would begin the process of realigning its generation and production activities, better known as vertical disintegration.

Central & South West Corp. of Dallas, Texas is a good example of a progressive electric utility that is already planning unbundling generation from production activities.

Central & South West Corp.'s Chief Executive Officer, E.R. Brooks, said the company is preparing for an unbundling of the industry into generation, transmission, distribution, and services segments by beginning now to realign its generation and production activities.

Other Structural Differences

Another very real, and often forgotten fact, is that electricity cannot be stored (with the exception of superconducting magnetic energy storage devices). There is no way on earth to harness electricity in a pipeline, for instance, to transport it to an end-user. As you would have guessed, substantial natural gas storage is available.

The NRRI also contends, as others do, that the potential for large stranded investments exist with retail competition in the electric power industry. Detail on stranded investments is discussed in chapter 4.

It has been estimated that stranded costs in the electric utility industry may be as much as 10 times more than that of the natural gas industry. On the gas side, there were very little (as compared to the electric utility industry) stranded investment issues—around $35 billion-with the advent of deregulation.

HOW ELECTRIC INDUSTRY DEREGULATION WILL IMPACT THE GAS INDUSTRY

Broadly stated, the most obvious aspect of the two industries connectedness is that they are both forms of energy. Being forms of energy make them both highly relevant in the overall scheme of energy marketing.

Although my discussion of the natural gas industry in this book will be in an abbreviated form, this in no way impacts on the relevancy of natural gas to energy marketing. My intent in this chapter is to show the significant causal connection between electricity and natural gas. There is a definite cause and effect relationship that exists between the two industries. What will or may be expected to happen in the natural gas industry as a result of electric utility deregulation, as this chapter will illustrate, is profound.

As some industry observers have noted, what will happen as a result of deregulation in the electric utility industry is that the natural gas industry will experience some very specific, marked effects.

Fuels Competition Likely to Be Exacerbated

Perhaps the most obvious, and maybe the most damning for the natural gas sector, is that electric utility industry deregulation will (most likely) pit electricity against natural gas (as well as alternative fuels) as an energy source in both the supply and end-use markets.

Electricity and natural gas already are in some competition with one another. However, opening up the electricity market to competitive forces most likely will place electricity in more direct competition with natural gas.

Electric industry deregulation also likely will lead to the desire on the part of the natural gas companies to become more competitive in the electricity sector. There is incentive to do so because a tendency exists for natural gas utilities to have the same customers as the electric utilities.

This heightened inter-fuel competitiveness scenario especially would hold true if deregulation in the electric utility industry leads to lower prices—which more than likely will happen at some point. The Federal Energy Regulatory Commission in FERC Order 888, the final restructuring ruling made by the federal agency on April 24, 1996, estimates that electric consumers nationally will save between $3.8 billion and $5.4 billion per year. Chapter 2 discusses this order in detail.

COMPETITIVE PRICE POSITION

A recent study conducted by R.J. Rudden Associates Inc. on the competitive price position of natural gas as a result of electric utility industry restructuring provides some valuable insight. The study focused on how much electricity rates would decline with deregulation, which markets would be affected by the decline, and the effect the decline would have on those gas end-use markets that compete directly with electricity.

FIGURE 3-1 COMPARISON OF ELECTRIC POWER AND NATURAL GAS INDUSTRIES

$$\text{CPCI} = \frac{\text{Competitive Electricity Price} - \text{Present Electricity Price}}{\text{Present Electricity Price} - \text{Present Gas Price}^{*}}$$

* The difference between the present electricity price and the present gas price is the "present natural gas advantage."

Source: Reprinted with permission by R.J. Rudden Associates, Inc., Copyright 1996.

From the study, an index was developed to express the potential change in electricity prices (in dollars per MMBtu) as a percentage of the price advantage that natural gas currently has over electricity. This index is referred to as the Competitive Price Change Index or CPCI. Computation of the CPCI is shown in Figure 3-1.

R.J. Rudden stated that a negative CPCI figure translates into the assumption that natural gas possibly can lose some of its current price advantage relative to electricity. A positive CPCI means that there is the possibility that gas could gain an increased price advantage over electricity.

Results showed that under full electric industry deregulation, natural gas has the potential to lose a significant portion of its price advantage during the winter in all North American Electric Reliability Council (NERC) regions and in all market segments, which includes

FIGURE 3-2 NORTH AMERICAN ELECTRIC RELIABILITY COUNCIL

ECAR
East Central Area Reliability Coordination Agreement
ERCOT
Electric Reliability Council of Texas
MAAC
Mid-Atlantic Area Council
MAIN
Mid-America Interconnected Network
MAPP
Mid-Continent Area Power Pool
NPCC
Northeast Power Coordinating Council
SERC
Southeastern Electric Reliability Council
SPP
Southwest Power Pool
WSCC
Western Systems Coordinating Council

Affiliate
ASCC
Alaska Systems Coordinating Council

Source: Reprinted with permission by the North American Electric Reliability Council, Copyright 1996.

residential, commercial, and industrial. The nine NERC regions are shown in Figure 3-2.

According to R.J. Rudden, this suggests that electricity has an opportunity to improve its market position within space heating markets. Natural gas' competitive price position, the study states, stands to decline by between 26% and 72% in the winter. The study found that on a regional, load-weighted, average basis, natural gas' competitive price advantage stands to decline by 48% in the residential market segment, 46% in the commercial market segment, and 41% in the industrial market segment.

Summer months tend to change the picture. According to the R.J.

FIGURE 3-3 DIFFERENCES BETWEEN FULLY COMPETITIVE ELECTRICITY PRICES AND PRESENT RATES (Expressed in 1994 cents/kWh)

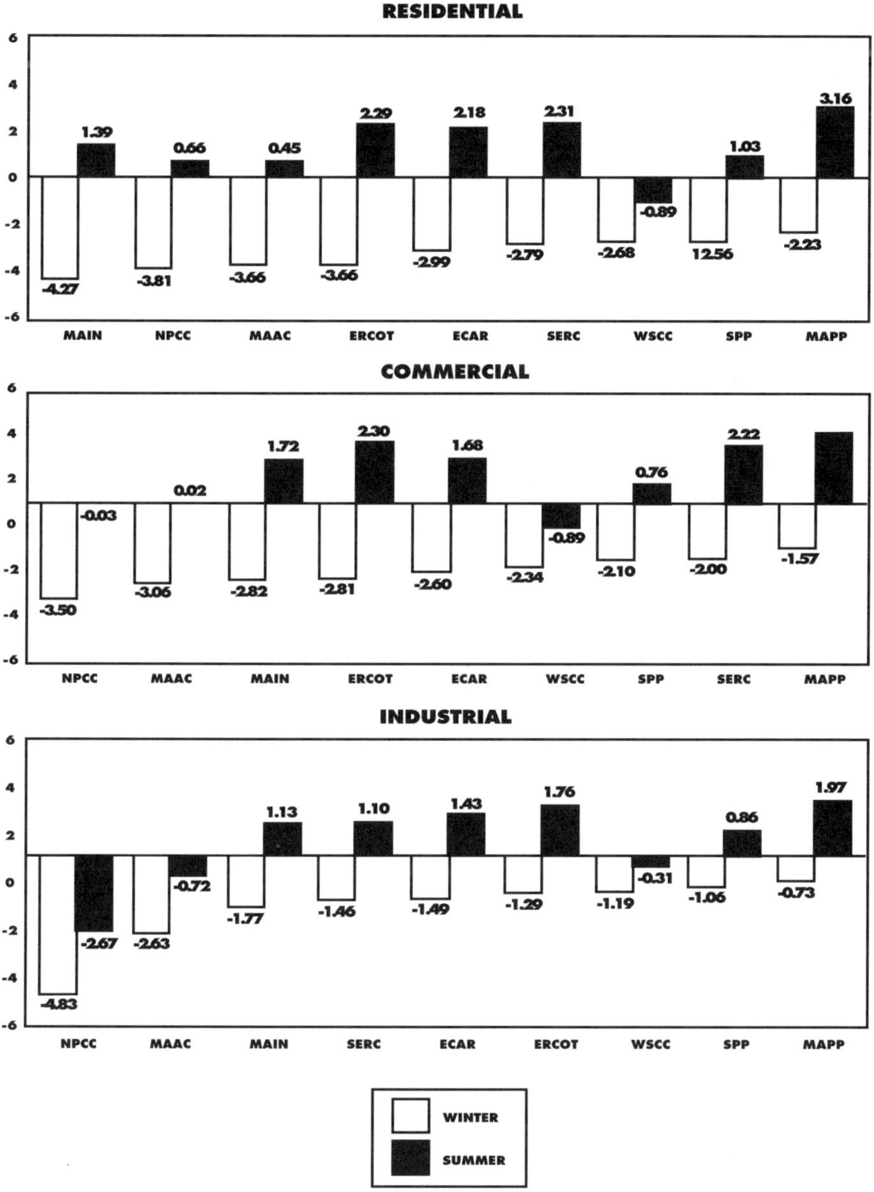

Source: Reprinted with permission by R.J. Rudden Associates, Inc., Copyright 1996.

FIGURE 3-4 COMPETITIVE PRICE CHANGE INDEX VALUES

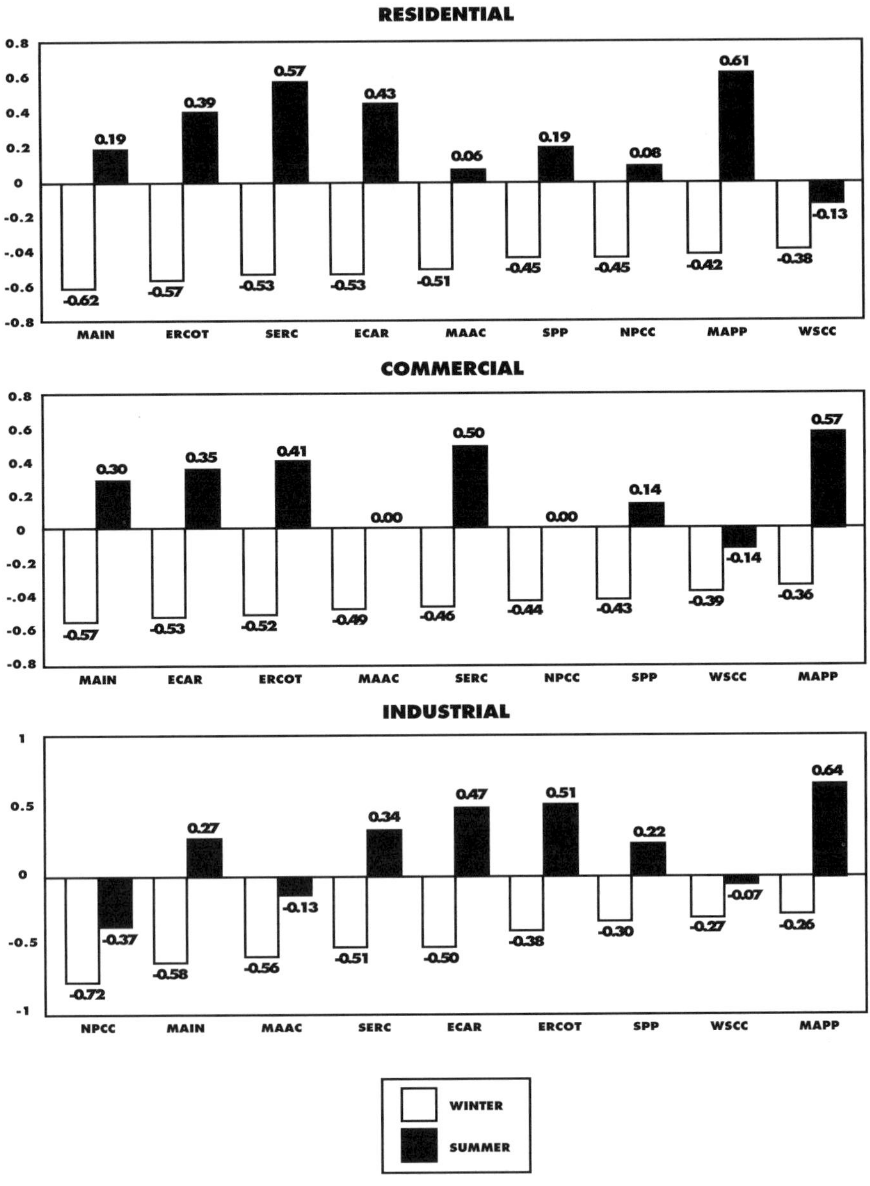

Source: Reprinted with permission by R.J. Rudden Associates, Inc., Copyright 1996.

Rudden study, gas' price advantage during the summer, for most of the market segments, will tend to improve. The study attributed the improvement to the fact that competitive electricity prices will actually track marginal costs more closely than regulated rates.

Further explanation offered by R.J. Rudden stated that because all NERC regions are summer-peaking and peak-related marginal costs are higher than off-peak costs, summer rates have a high probability of increasing relative to winter rates.

Specifically, the study found that on a regional, load-weighted, average basis, the summertime competitive price advantage of natural gas for the residential segment could improve by 24%. For the commercial market, gas' price advantage could improve by 22%. The industrial market could see an improvement of 26%.

These results, according to the study, suggest that in certain end-use markets in the summertime, such as space cooling, gas has the potential of being able to compete effectively with electricity.

According to R.J. Rudden, its computational process gave birth to two sets of results: the difference between competitive electricity prices and present electricity prices, and CPCI values. Figures 3-3 and 3-4 represent each of these two sets of results.

R.J. Rudden stated that its study didn't account for or address the extent of electric price reductions some larger customers might achieve by negotiating shorter-term and/or interruptible service agreements, or by availing themselves of "special contract" rates-such as load retention and economic development rates.

The study concluded that the previously described special arrangements that can be struck between an industrial customer and an electric utility pose a sizeable risk to the gas industry.

GAS INDUSTRY FACES CHALLENGES

With the looming deregulation of the electric utility industry, the natural gas industry has braced itself. This is largely due to the fact that some of the challenges the gas industry faces have already hit. For example, the many uncertainties that are following the pending deregulation of the electric power sector have already begun having a downward effect on natural gas demand.

In addition, the electric utility industry will be faced with—by conservative measures—at least $135 billion in stranded investments, according to Moody's Investors Service. Recovery may actually cause the market value of electric generating assets to plummet. This would have a domino effect on the gas industry: it would further increase the problem of excess generating capacity due to reduced gas demand.

Electric utility industry deregulation will pose significant challenges, opportunities for the natural gas industry.

In summation, deregulation in the electric industry sector means for the natural gas sector that it must be willing to be flexible, open, and accepting of the challenges it will face in the future.

OPPORTUNITIES FOR THE GAS INDUSTRY

Opportunity for the gas side also exists, however, with the advent of electric power industry deregulation. A diverse power generation market, for instance, will be opened up to natural gas as a result of electric industry restructuring.

In addition, the electric utility industry in the future should find that gas-fired plants generally have lower operating and maintenance expenses, as opposed to traditional coal and nuclear-fired generation sources. In the future, then, there is an excellent chance that more plant constructions will be gas-fired, due to their less risky nature and lower capital cost requirements.

It should be noted, however, that in the short-term, capital demands and the underutilization of current plants by electric utilities will serve to dampen new natural gas capacity additions.

It becomes clear then that the electric utility industry and natural gas industry exert a significant amount of influence on one another. This influence most assuredly will be exacerbated by the introduction of competition in the electric sector.

Deregulation presents both challenges and opportunities for not only the gas industry, but the electric power industry as well. It will not be surprising in the future, as it is already happening currently, that the gas industry and the electric power industry will increase their efforts to team up through special alliances and mergers to tackle the national energy market together. There is strength in numbers. These special alliances will be more fully developed in chapter 6.

CHAPTER 4

RETAIL WHEELING

The term retail wheeling has only fairly recently become a so-called *buzzword* within the industry as players in the electric utility industry have begun the trek toward a competitive electric power marketplace.

Retail wheeling has been in existence now for several years, however, it only has been permitted in very limited circumstances. States, as well as individual utilities, and even FERC have permitted the practice in the past but have only done so in special arrangements.

It must be pointed out early on in this chapter that retail wheeling on a wide-scale basis, although never been done before, is on a fast-track to becoming the standard rather than the exception.

WHAT IS WHEELING?

Wheeling, in general, is the use of transmission facilities of one network to deliver power of and for another entity or entities. There are essentially two very broad and different forms wheeling can take. They are retail wheeling and wholesale wheeling.

Retail wheeling is the transmission of power to an individual customer for a generator of electricity other than the host utility. It also is referred to as retail access or direct access.

Wholesale wheeling, on the other hand, occurs when the buyer of power resells the wheeled power to retail customers. Chapter 5 addresses wholesale wheeling in more detail.

THE FORMAL INTRODUCTION OF RETAIL WHEELING

The Energy Policy Act of 1992 (EPAct) first formally introduced the electric utility industry to the term retail wheeling. This would mark the *formal* beginnings of retail wheeling. Passage of the act also would focus a great deal of attention on regulatory reform as a key driver in the industry's transition to open price competition.

Retail wheeling: the transmission of power to an individual customer for a generator of electricity other than the host utility. It also is referred to as retail access or direct access.

While EPAct triggered a variety of reform proposals for the electric power industry, the industry's transition also is rooted firmly in changing market conditions. Nonetheless, the act most assuredly was the historical event that touched off the wave of restructuring proposals currently underway in the states.

Interestingly, even though EPAct marked the first time the utility industry was formally presented with the term *retail wheeling*, the act was painfully vague about the nitty gritty specifics of the practice.

However, there was one aspect tied to retail wheeling that was made crystal clear in the act. Specifically, Sections 721 through 726 of EPAct expressly prohibit the Federal Energy Regulatory Commission (FERC) from mandating the practice. Although there is some debate, many in the industry believe that EPAct, because it doesn't forbid states from permitting retail wheeling, leaves to the states the last word on whether or not retail wheeling will be permitted in the individual states.

NATIONWIDE RETAIL WHEELING SOUGHT

There has been a recent movement on the part of several congressmen to work for a federal retail wheeling statute. Congressmen involved thus far include Sen. J. Bennett Johnston (D-Louisiana), Rep. Dan Schaefer (R-Colorado), and Rep. Edward Markey (D-Massachusetts).

Sen. Johnston was the first congressmen to introduce comprehensive electric utility deregulation bill. That Bill, S. 1526, the "Electricity Competition Act of 1996," mandates nationwide retail wheeling among electric energy suppliers by the year 2010. Johnston's bill also provides for full recovery of stranded costs.

There appears to be no major legal constraints to Johnston's bill because it does not attempt in any way to pre-empt the states from ordering retail wheeling. Johnston's bill and the other bills that have been introduced basically are working to ensure that competition via retail wheeling is *standard* among *all* states.

There are six main elements to Johnston's bill. They are:

- retail access
- stranded costs
- shared federal and state responsibility
- a timetable for the transition to competition
- Public Utility Regulatory Policy Act (PURPA) reform and Public Utility Holding Company Act (PUHCA) repeal
- Nuclear power plant decommissioning and cost recovery

Avoiding the "Patchwork Quilt"

Stated again, the proposed federal legislation does not attempt to usurp state authority as it relates to the enactment of retail wheeling. What the legislation does attempt to do, however, is to present the industry (the states, specifically) with some general guidance on retail wheeling.

Johnston and others have stated that a federal approach to mandating retail wheeling nationwide would serve to do away with the current, and expected future, patchwork quilt approach.

This patchwork quilt approach makes it complicated not only to regulators but also to industry observers. Trying to keep up with retail wheeling on a state-by-state basis is cumbersome and ever-changing. As such, I will not attempt here to do a rundown listing of each state; it would change the very second this book hit the market. I have, however, provided in chapter 2 a skeletal look at what California is doing in this area, because it has been an industry leader with respect to retail wheeling, and, consequently, the most outspoken.

STATES PLACE RETAIL WHEELING AT TOP OF AGENDAS

At present, nearly every state, every utility, and every public utility commission has investigated retail wheeling, sometimes even if there is some opposition or otherwise hesitant feelings about permitting the practice.

This is largely because there is the realization that retail wheeling is *inevitable* at some point. Beginning now the process of investigating retail wheeling is a definite and recurring agenda item in most states and at most utilities and will continue to be well into the foreseeable future.

PILOT PROGRAMS

The evidence that major stakeholders are taking very seriously how they approach the issue of retail wheeling can be found in the fact that there are a handful of pilot programs that are either under way or are in the planning stages. Retail wheeling pilot programs have three common elements. They are:

- Made available to a specified number of participants
- Offered to a specified group of customers
- Available only for a specified amount of time

The function of these particular pilot programs can be most useful in determining customer reactions to the issue of choosing their own energy provider. Pilots also give utilities a dose of much-needed experience with retail wheeling, if only in a limited way.

It is extremely important in any pilot program that the results, or the lack of, whichever the case may be, can be measured. So, many utilities employ (or will employ) the use of specially-set milestones or goals as a method of determining how purposeful the pilot is.

For example, if the goal is to determine if customers or end-users can find cheaper sources of electric power, then the utility may log which utilities their customers are choosing, the prices of the competing provider, and other similar factors.

These findings, taken as a whole, can serve to help utilities come up with ways to cut their costs through innovative methods (which might include changing generating practices, for example) to help in bolstering their competitive position with customers. For it is largely through lowered operating costs that utilities are able to keep their competitive foothold—and retain their current customers. The desire to retain current customers consequently also serves as a major driver in a utility's decision to institute a retail wheeling pilot program.

THE PLAYERS: WHO'S FOR AND WHO'S AGAINST RETAIL WHEELING

A chasm has been building for some time now between those in the industry who oppose retail wheeling and its effects and those who favor the practice and the expected efficiencies it may produce. It is not difficult to conclude that those who have not accepted the reality that retail wheeling will someday become a reality are merely deluding themselves. This can only serve to injure the staunch opponents' competitive positioning in the new deregulated order of the industry.

It is an over-generalization to say that the list I am about to provide of who is against and who is for retail wheeling is all-encompassing or that it applies to all in a certain class.

Retail Wheeling Opponents

That said, most usually, those classifications of industry players most likely to oppose retail wheeling include *some* utilities, small consumer groups including conservationists and environmentalists, the financial community, and others. It can be said that each of the groups feels that the interests they serve will somehow be damaged should retail wheeling become the rule rather than the exception.

Retail Wheeling Proponents

Those favoring retail wheeling include industrial consumers, nonutility power producers, some economists, as well as others. It must be noted that, while there may be some utilities that remain opposed to retail wheeling, there are a fair number of them who are not.

Where utilities are opposed to retail wheeling it generally holds true that utility rates are already, for example, in line with regional or area rates or with the national average. However, there is exception to this as well, as there will be throughout this discussion on the proponents and the opponents.

A good example of an exception is the state of Oklahoma. Despite recording some of the lowest electricity rates in the United States, the state is not ignoring the retail wheeling issue. In fact, there is a retail wheeling "shell" bill awaiting further input from industry stakeholders.

Conversely, it appears as though those states with exceptionally high electricity rates are ready for retail wheeling. These states—California is a good example—generally advocate implementing retail wheeling as soon as practicably possible.

On the other hand, states which oppose the practice would rather see (if they have to see) retail wheeling phased in incrementally. Caution and a go-slow approach are both characteristic of opponents of retail wheeling in addition to being characteristic of those who are still as yet unsure of the positive merits of the practice.

WHY RETAIL WHEELING NOW?

The potential for improved economic efficiency is the main motivator behind the current debates and activities surrounding retail wheeling.

According to the National Regulatory Research Institute (NRRI), there are several factors behind the heightened interest in retail wheeling. They include:

- Large price differentials between utilities

- High electricity prices relative to the costs of new generation facilities

- A recent emphasis on economic development and new jobs has led some industry stakeholders (including state Legislatures, public utility commissions and industrial groups) to push for more competitive electricity prices.

- Industrial customers feel that utility-funded demand-side-management (DSM) programs have caused electricity prices to increase. They also argue that most of the benefits from DSM programs have mostly benefited nonindustrial customers.

- The general belief of many stakeholders that more competition in the electric power industry would have positive repercussions.

POLICY QUESTIONS

In addition to offering several factors that are main driving forces behind retail wheeling, the NRRI offers 12 policy questions, as depicted in Table 4-1. These and other questions are currently being debated by the major stakeholders in the electric utility industry, including electric utilities, state public utility commissions, state Legislators, end-users and others.

TABLE 4-1 POLICY QUESTIONS RELATING TO RETAIL WHEELING

1. What would be the meaning of "obligation to serve" in a retail wheeling environment?
2. How would services be unbundled and priced to accommodate retail wheeling?
3. How would the structure of the electric power industry change?
4. What services would continue to be regulated? How should they be regulated?
5. How could regulation and competition coexist?
6. How could core customers be protected?
7. What would be the responsibility of departing customers for the local utilitie's fixed costs?
8. Should the "regulatory compact" be redefined and, if so, how?
9. When and under what conditions should retail wheeling be allowed?
10. How would conflicts between core and noncore customers be resolved
11. How would pricing for different electricity services change?
12. How would utility-financed DSM activities and the IRP process be affected?

Source: Reprinted with permission by the National Regulatory Research Institute, Copyright 1994.

As the table shows, such open-ended questions as, "How would the structure of the electric power industry change?" and "When and under what conditions should retail wheeling be allowed?" will ultimately be answered (perhaps in several different ways, however) as the deregulation of the industry continues to shake out.

COSTS AND BENEFITS

There are many costs and benefits associated with retail wheeling. In a very real sense, these costs and benefits are difficult to define exactly, however, there are some quantifiable ones. They are listed in Table 4-2.

TABLE 4-2 POTENTIAL BENEFITS AND COSTS OF RETAIL WHEELING

Benefits	Costs
1. More efficient utility pricing	1. Lower electric power system reliability and stability
2. More effecient utility operations and investments	2. Uneconomic bypass under existing retail pricing procedures
3. More appropriate "regulatory compact"	3. Stranded-investment costs
4. More efficient industry structure	4. Large distributional effect
5. Reduced price differentials among electric utilities	5. Lost economies of scope
6. Stronger U.S. economy	6. Incremental costs for upgrading or expanding transmission network to accommodate retail wheeling
	7. Jurisdictional disputes
	8. Higher prices to core customers
	9. Discriminatory pricing
	10. Breakdown of "regulatory compact"
	11. Abolition of IRP process

Source: Reprinted with permission by the National Regulatory Research Institute, Copyright 1994.

It can be said that the most vital cost and the one receiving the most attention has been that of stranded costs and recovery. Sometimes called stranded investments, stranded costs are capital costs, such as for power plant construction, that utilities incurred based on a stable customer base. These capital expenses are strandable in a competitive environment because they could go unrecovered if power customers chose another supplier other than the original host supplier.

COST RECOVERY

Shortly before this book went to press, The Federal Energy Regulatory Commission issued its final rule, FERC Order 888, on the restructuring of the electric utility industry. Among many other concerns, the final rule addressed the recovery of wholesale stranded costs. FERC Rule 888, to the satisfaction of many, allows utilities to fully recover these costs, with a few exceptions.

FERC Rule 888 does permit utilities to recover stranded costs from their departing customers. These so-called "exit fees" would consist of special charges levied on customers who leave the native utility to take advantage of less expensive power from other power suppliers. FERC Order 888 is covered more fully in chapter 2.

CHAPTER 5

WHOLESALE WHEELING

Wheeling refers to the transmission of electricity from a seller of the power to a buyer of the power over the transmission lines of one or more other entities, including utilities and government facilities.

Wholesale wheeling, in particular, occurs when the buyer of the power resells the wheeled power to retail customers.

The practice has been in existence for several years, but it wasn't until the passage of the Energy Policy Act of 1992 (EPAct) that it received more attention through the opening up of new opportunities. These were opportunities that would not only continue allowing utilities to wheel power among themselves, but would open the door wide open for nontraditional players to try their hand at wholesale wheeling. Those opportunities, plus other wheeling issues, will be presented in this chapter.

AN HISTORICAL BACKGROUND

A definition (provided earlier) along with the terms, requirements, and regulatory authority of wholesale wheeling transactions can be found in Section 201 of the Federal Power Act (FPA).

The FPA was put into motion in 1920. There have been several amendments made to the act over the years. Among them was an amendment which grants FERC the regulatory authority over generation and transmission of electricity that is in interstate commerce. I will not attempt here to go into all the minute details of how and when FERC regulates wholesale wheeling. It does in most cases, but as in all activities in the energy industry, there are exceptions.

Another amendment was one that permitted a wholesale electric generator to ask FERC to order a transmitting utility to provide transmission service. Yet another amendment set up protocols for "good faith" wheeling requirements. These amendments, which were brought about through the passage of the Energy Policy Act of 1992, and their implications, are discussed in detail below.

THE ENERGY POLICY ACT AND ITS IMPLICATIONS

As stated in previous chapters, EPAct is seen by many in the power industry as a turning point in the regulation of public utilities in several different ways. This view of EPAct certainly holds true to form as it relates directly to the practice and regulation of wholesale wheeling activity, and as will be seen in this chapter, to sales opportunities.

The FPA was impacted by the passage of EPAct in two very significant ways. First, under Section 211(a) of FPA, which was amended by EPAct, all entities generating electricity for wholesale (with some exceptions, of course) were granted the ability to request an order from FERC for a transmitting utility to provide transmission service.

The entity requesting wheeling orders has to meet certain requirements and the wheeling arrangement must be in the public interest. Public interest concerns are addressed in Section 212 of the FPA. In most cases, utilities are willing to provide wheeling services so long as the above (and other) requirements are met. And, as will be discussed later in this chapter, independent power producers (including cogenerators), barring the inability to effect an agreement with a utility, can seek FERC intervention.

The other revision to FPA via passage of EPAct was the establishment of special so-called "good faith" requests for wheeling services. Good faith requests, according to FERC, must include such information as who the purchaser is, the type of service that is being requested, the proposed dates for beginning and ending wheeling service, and whether the service will be firm or nonfirm.

There are other pieces of information required from a potential purchaser to make the deal a so-called good faith request, but the ones listed here will give you an idea of how FERC defines a good faith request for wheeling services.

In addition, the *seller* must meet certain good faith request requirements. For example, the seller is required to let the purchaser know within 10 days that it has received the request. There are other requirements and this is just one of them.

Another amendment to FPA made by EPAct was one in which FERC mandates that transmitting utilities annually submit information relating to capacity and constraints. This is to help keep transmission customers and regulatory bodies abreast of the activities at the utility.

THE ROLE OF INDEPENDENT POWER PRODUCERS

As stated earlier, EPAct removed almost all barriers that existed prior to its passage as they related to competition. Chapter 2 goes into even greater detail about EPAct and how it made important and far-reaching amendments to the Public Utility Regulatory Policies Act of 1978. At base, EPAct expanded and improved the ability of competition to pervade the industry.

This chapter will emphasize how EPAct was pivotal in opening the door for nontraditional entities to become more actively competitive. This allowed such entities as Independent Power Producers (IPPs) and cogenerators (discussed later in this chapter) to begin or step-up their activities in developing, building, owning, and operating power generating facilities.

Independent Power Producers, having experienced more than a decade of PURPA and its requirements, benefitted greatly from the amendments contained in EPAct. Over the years, IPPs truly have shown the power industry that they can compete with utility generation on a cost basis.

The fact that IPPs can compete with utilities is evidenced by the fact that through August 1994, U.S. electric utilities issued requests for proposals (RFPs) from IPPs to supply over 35,400MW of new capacity. The RFPs were answered by bids that totaled nearly 420,000MW.

Too, those utilities that purchase power from the wide array of independent power companies they have to choose from have the ability to determine which technology or technologies, fuel(s), permitting, siting, financing, dispatchability, and pricing characteristics would best fit their overall needs and the needs of their customers.

IPPs are projected to account for over half or 45,000MW of new U.S. generating capacity in the next decade. A good deal of attention will be paid them in this chapter. Independent power definitely is taking and will take a large chunk of the attention in the industry over the next decade and beyond.

It must be pointed out that IPPs are not really new at all to the competitive energy environment. In fact, IPPs have been a significant driving force in the evolution of competitive power markets for more than a decade.

According to the National Independent Energy Producers (NIEP), IPPs supply approximately 50,000MW of capacity or 7% of the nation's power. Over $500 billion has been invested by independent power companies in more than 3,200 power projects across the country. These projects, according to NIEP, account for around $10 billion in annual electric sales.

During the period 1990-1993, IPPs were responsible for 53% of all new capacity in the United States and are projected to continue providing a large percentage of power into the future. According to NIEP, more than 1,400 independent power projects are under development. Completion of these projects will require IPPs to commit up to $10 billion in investment dollars. The accompanying sidebar, "The Project Financing Model," explains detail surrounding IPP financing structures and sources.

THE PROJECT FINANCING MODEL

Independent power producers (IPPs) have relied on a variety of financing structures and sources to fund project development and construction. The most common method has been asset-based project financing, in which the facility developers pledge the asset and a portion of the project's revenues to lenders.

For a typical IPP project, a developer will formulate a plan to build a generating facility and will secure a long-term contract to sell power to an electric utility. The developer will approach one or more financial institutions—usually banks or institutional investors, such as insurance companies—to lend capital for project development and construction. Depending on the strength of the developer and its project proposal, the lender will provide debt financing for a large share of the project costs, perhaps 80%. The remaining 20% will usually be provided by the developer in the form of equity, or the developer will bring in other equity investors.

Financing IPP facilities can be seen as similar to home mortgage lending. A home buyer will provide a down payment (equity) of perhaps 10% of the house's selling price. A bank will provide a mortgage (debt) for the remaining 90% of the cost. The homeowner promises part or all of his or her income (revenues) to pay off the loan, and offers the house (the asset) as collateral.

Independent power developers are using variations on this model to fund their projects. For example, a developer may obtain financing for part or all of a project's costs by issuing public debt bonds. Or it may issue bonds for sale to certain institutional investors, such as insurance companies, mutual funds, and pension funds. Or, the company may issue a public offering of stock, and use proceeds from the sale of shares to fund development.

With an estimated $50 billion of IPP investments in generating facilities, financing IPPs has become a large business. Dozens of financial firms, lenders, investment banking firms, and financial consultants provide specialized services for financing independent power facilities.

Reprinted with permission
from the National Independent Energy Producers

COGENERATORS

Cogeneration facilities fall under the umbrella of an independent power producer and are a class of non-utility generators (NUGs) that are characterized by their close physical proximity to the user of the electricity they generate and are directly sited at their thermal host. Cogeneration, by definition, is the production of both electric and thermal power from an energy source.

As stated and explained in earlier chapters, the Public Utility Regulatory Policies Act (PURPA) put into place a new statute that placed utility and nonutility generation on equal footing with respect to economic considerations. The act gave an important boost to the construction of cogeneration facilities through allowing industries to sell power to utilities so long as they matched the utility's cost of new generation capacity installation.

A look at how a simple wholesale cogenerated wheeling system operates is shown in Figure 5-1. In addition, Table 5-1 lists some of the most important technical options that should be considered before entering into a cogen project.

TABLE 5-1 TECHNICAL WHEELING OPTIONS

Without Wheeling	With Wheeling
Energy conservation at host	Larger cogeneration facility
Future plant expansion	Peaking power production
Design alternatives	Supplemental fuels
Disposition of electricity	Wheeling at contract expiration
Fuel alternatives	QF versus EWG status

Source: Reprinted with permission by PennWell Publishing Company, Copyright 1995.

FIGURE 5-1 WHOLESALE WHEELING COGENERATED POWER SYSTEM

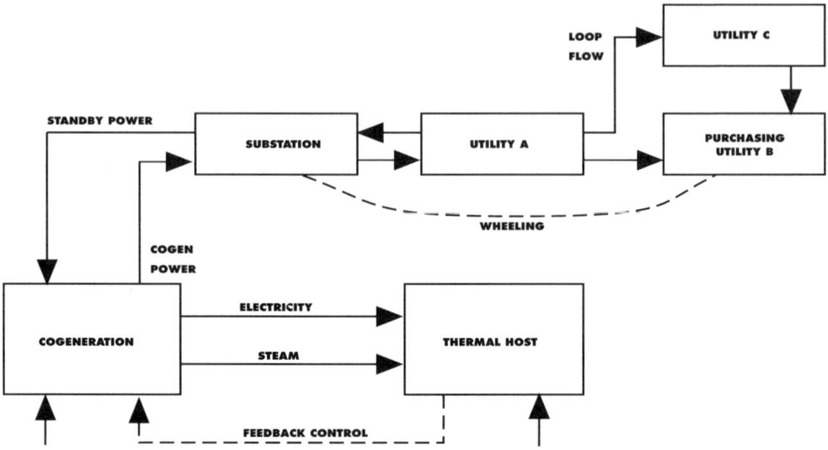

Source: Reprinted with permission by PennWell Publishing Co., Copyright 1995.

NON-UTILITY GENERATORS

Also falling under the wide umbrella of independent power facilities are non-utility generators or NUGs. As an independent power source, like cogeneration facilities, NUGs are increasing their share of power generation.

According to a 1995 Parks Associates report entitled, "Electric Utilities: Regulatory, Financial & Social Forces for Industry Change," NUGs often produce electricity much more cheaply than other types of generators due to their ability to use the *free* heat from the production process.

The Parks report stated that EPAct in the coming years can be expected to provide NUGs with improved flexibility in choosing the size, fuel, and location of new plants.

This added bonus that may be granted to NUGs is actually quite profound because, as Parks Associates states, this provision will allow NUGs to sell to specific wholesale purchasers who are shopping for the least expensive power supply from a wide variety of choices in a much larger geographic area. The potential ramifications to this are potentially significant, because it could be that in the future cogenerators could be wheeling to retail customers.

Parks Associates stated that due to a variety of conditions brought about especially through PURPA and EPAct, utilities are purchasing new power capacity through competitive bidding, where NUG power contracts are awarded through a competitive bidding process among potential suppliers. In addition, the report states, because utilities no longer have a generation monopoly, NUGs have been increasing their share of power generation. This is shown in Figure 5-2.

FIGURE 5-2 UTILITY VS. NONUTILITY GENERATION ADDITIONS

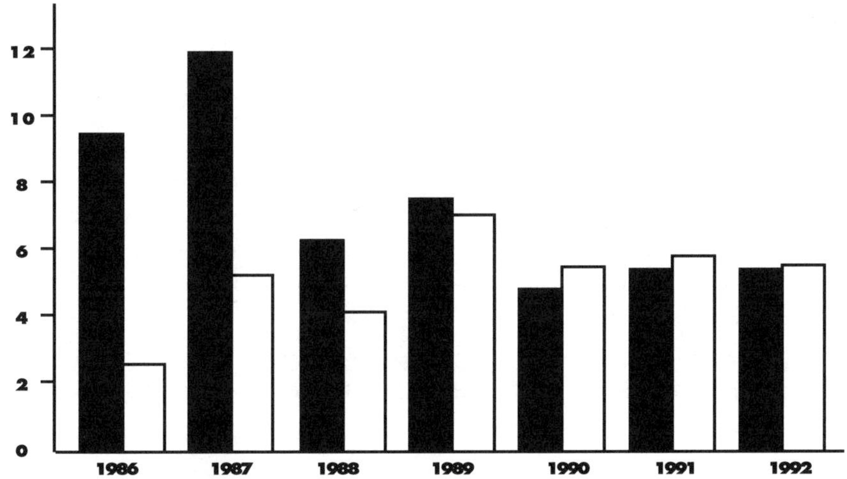

Source: Reprinted with permission by Parks Associates, Copyright 1995.

NUG Agreement Buyouts

According to a study by the Washington International Energy Group (WIEG), which polled electric utilities and independent power producers on a variety of industry issues, respondents reported very little action in restructuring their power purchase agreements (PPAs) with nonutility generators.

The study, "The 1996 Electric Industry Outlook," found that only 18% of respondents indicated that their companies have bought out existing PPAs for new projects, while an overwhelming 77% have not.

The study also discovered that, among the respondents, 45% say their companies do *not* plan to buy out or renegotiate any existing PPAs, while only 32% are planning to do so.

As shown in Figure 5-3, while significantly more respondents—76% compared to 62% last year—believe that renegotiations and buyouts will increase, the WIEG study found that many utilities have no *immediate* plans to do so.

MARKETING IMPLICATIONS

As stated earlier, passage of EPAct afforded cogenerators increased sales opportunities. EPAct actually had the effect of opening up cogen sales opportunities to off-site or distant utilities through wheeling. And, as will be shown below, cogenerators potentially will have a significant impact on energy marketing activities and vice-versa.

Energy Marketer Function

Because energy marketing is all about providing energy solutions to customers (mostly industrial at the present time), it stands to reason that energy marketers can be a big help to cogenerators. And, likewise, cogenerators can play a vital role in offering the marketers just one more cost-effective alternative to offer their customers.

Energy marketers can aid cogenerators in that they have the resources to determine where and what the market might be for cogenerated power. Marketers will serve this function, but as well, they can facilitate access to transmission lines.

FIGURE 5-3 PERCENTAGE OF RESPONDENTS WHO BELIEVE BUY-OUTS OF NUG AGREEMENTS WILL INCREASE

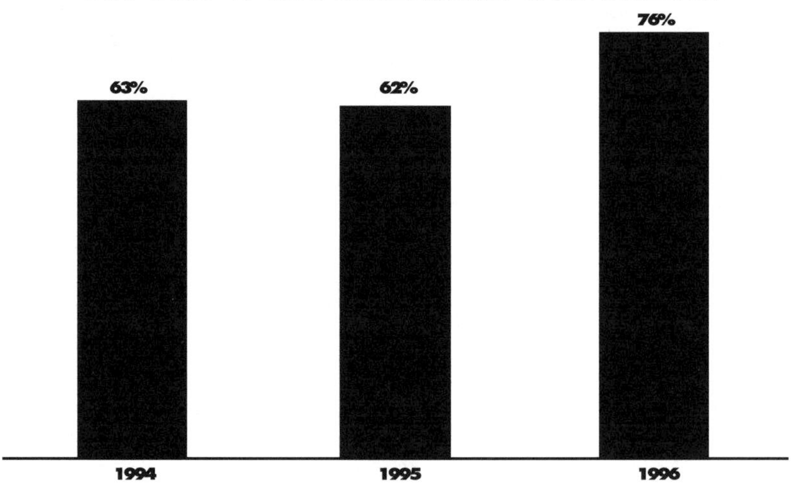

Nonutility Generator Agreement Buyouts (Reprinted with permission by the Washington International Energy Group, Copyright 1996.)

As stated earlier in this chapter, cogenerators as NUGs can produce electricity cheaply, and they have several competitive advantages over utilities in that they have the needed experience at their disposal to build the type of facilities that lead to the lower-cost power. The fact that NUGs in general have considerable clout in the industry is evidenced by the fact that of the almost 13,000 MW of U.S. electricity projects that are awarded through the competitive bidding process, NUGs won 85% of the projects; utilities won 15%.

Of interest, and to serve as an example of how cogenerators are entering into power sales agreements with large energy users and are effecting considerable efficiencies, Figure 5-4 shows a photo of the King City cogen facility. Calpine Corporation, a national power marketer, has entered into a lease with BAF Energy, a California limited partnership for this 120MW gas-fired cogeneration power plant.

FIGURE 5-4 KING CITY COGENERATION FACILITY

Source: Permission to reproduce granted by Calpine Corporation.

Since start-up in 1989, the plant has operated at an average 97% availability factor. Electricity will be sold to Pacific Gas & Electric Co. under the terms of this long-term power sales agreement. The steam will be sold to a vegetable processing facility, which is owned by an affiliate of BAF Energy.

The transaction strengthens the purchasing power of Calpine's fuels division by adding a fuel demand of 4 to 5 billion cubic feet. What's more, the transaction increases the gas-fired projects Calpine operates to over 400MW and its total net interests in power generating facilities to 863MW.

TRANSMISSION ACCESS

Some concern exists over whether or not utilities that own transmission will hamper their new competitors (i.e., cogenerators). Utilities have a vested interest in off-system, short-term power purchases and sales that reduce costs for their customers.

Because cogenerators generally have long-term contracts under firm prices, there has been some concern in the past that utilities may, over the long haul, hamper transmission access.

FERC Order 888

To address these concerns, FERC issued a final ruling on open access, FERC Order 888. Details are provided in chapter 2. Essentially, Order 888 provides for open and fair electric transmission services by public utilities.

In particular, Order 888 sets out a single pro forma tariff, which describes the minimum terms and conditions of service that public utilities must offer to bring about non-discriminatory open access transmission services to all. Under this pro forma tariff, all public utilities that own, control, or operate interstate transmission facilities must offer comparable service to all others that request it.

CHAPTER 6

ALLIANCES AND SPECIAL MARKETING RELATIONSHIPS

Just as competition has spurred increased opportunity for nontraditional players in the energy industry, so too has competition opened up within the industry a very nontraditional trend. This new and unprecedented trend revolves around participants in the industry working together to provide energy solutions for the benefit of customers.

In the past, it was rare to find energy companies, both electric and gas, actually working in concert to help each other find ways to serve customers better. Also rare was the now-current push to *please* customers. Customer satisfaction has become a tantamount concern as deregulation begins its quick descent onto the electric power industry.

It no longer is kosher to "go it alone." Energy companies are finding that working together gives them strength—both in sheer size and in capital resources—and allows them to better serve their

The main driving force behind utilities' concerns over improving and enhancing customer service is deregulation.

customers. The main driving force behind utilities' concerns over improving and enhancing customer service is deregulation.

Today, price is king. The time is ripe for utilities to work together with other energy providers (i.e., gas) as well as *directly* with their customers to offer the lowest-cost power available to end-users. Retaining customers and perhaps drawing in a few more in the process will be the difference between remaining in business and going out of business.

Utilities are finally beginning to learn that their customers are the lifeblood of their businesses. They are taking heed. This chapter will present these new relationships and how they are taking shape as the electricity industry deregulates.

ENERGY SERVICE COMPANIES

Energy service companies or ESCOs have just recently become all the rage in the power industry. ESCOs present a very unique strategy companies can employ to get a foothold in the new competitive environment.

ESCOs are companies that are formed for the express purpose of satisfying customers. End-users become the driver for whatever ESCOs may decide to offer. ESCOs offer so-called value-added services—services that add some value for the end-user.

ESCOs present a very unique strategy companies can employ to get a foothold in the new competitive environment.

This value most often comes in the form of products and services that are of use to the particular desires and needs of both individual customers and classes of customers, such as commercial or industrial.

Value is also obviously found in the energy itself. It follows that ESCOs have as a main mission that of determining the form or forms of energy that would satisfy a particular customer's needs. In essence, ESCOs become a customer's "waiter" or "waitress." ESCOs will find out what the customer wants (the "order") and then do everything in their power to deliver (fulfill the patron's "order").

Most ESCOs will be, if they aren't already, involved in four different types of value-added activities. They are:

1. Products and services
2. Financial planning and consultation
3. Delivery options and efficiencies management
4. Commodities management

Although the formation of ESCOs is fairly recent, there already are a fair number of utilities that are teaming up with other utilities, both electric and gas, and with the end-users (customers) themselves. They tackle this business of providing value-added products and services to their customers.

In addition, as first brought out in chapter 3, special alliances not only are being formed within the electric utility industry but also between the electric and gas industries. As discussed later in this chapter there has even been a merger planned between a major electric utility and major gas company. The intent with this merger, like all the other alliances (including ESCOs) is to become a so-called "one-stop shop" for customer energy solutions.

ESCO FORMATION EXAMPLES

As stated earlier, there are a great many examples of utilities and other entities that have formed ESCOs. This chapter describes three in detail: Green Mountain Energy Partners, Lykes-Duke/Louis Dreyfus Ltd., and UtiliCorp United's EnergyOne℠.

Green Mountain Energy Partners

Green Mountain Energy Partners, which still must gain the necessary regulatory approvals, was formed by subsidiaries of each of the following: Green Mountain Power Corporation, Hydro-Quebec, Consolidated Natural Gas Co., and Noverco Inc.

Green Mountain Energy Partners will compete for energy sales to all classes of retail customers. This new company has targeted the New Hampshire electricity pilot program as the first market in which it will compete for retail customers. (The New Hampshire pilot is the first time that electric service will be opened to competition in New England).

Green Mountain Power is an investor-owned electric utility in Vermont with long-standing ties to Hydro-Quebec, including purchased-power agreements and a research-and-development agreement. Green Mountain Power's primary responsibility in the new partnership will center around marketing and retail sales.

Consolidated Natural Gas is participating in the enterprise through its CNG Energy Services Corporation subsidiary. Based in Pittsburgh, CNG Energy Services provides electricity, natural gas, other energy forms, and related services to customers across North America. Through an affiliate, CNG Energy Services is the fourth-largest wholesale power marketer in the United States, with sales of nearly 2 million MWhr in 1995. Its 1995 operating revenues are at $3.3 billion, making it one of the largest integrated natural gas companies in the United States.

Hydro-Quebec is one of the world's largest electric utilities and is owned by the Province of Quebec. It has electricity production capability, 97% of which is renewable-resource hydropower, which is 1.5 times the entire electric load for New England. Its revenues in 1995 were $5.3 billion.

Noverco is a Quebec energy holding company among whose properties are Gaz Metropolitan Limited Partnership, Quebec's natural gas distribution company and Vermont Gas Systems (Vermont's natural gas distributor). Noverco's operating subsidiaries had revenues of nearly $1.4 billion in 1995.

All four companies will continue their separate alliances involved in wholesale energy sales in the United States. Though a merger is not imminent among these companies, one thing for sure is that the combined market strength these companies will be able to share will be a highly valued asset—for the companies as well as for their customers.

Lykes-Duke/Louis Dreyfus Ltd.

Another example is that of a new company formed by Duke/Louis Dreyfus LLC and Lykes Energy Inc. The new company, Lykes-Duke/Louis Dreyfus Ltd., will provide energy-related services to both wholesale and retail customers in Florida. The following are among the activities the new entity will engage in:

- Electric and fuels management,
- Generation facility ownership or leasing and asset operations,
- Energy systems engineering and design, and
- Fuels procurement and management.

Duke/Louis Dreyfus and Lykes Energy plan on sharing in the ownership, operating costs, revenues, and the profits of the new entity. The new ESCO will require Federal Energy Regulatory Commission approval to engage in power marketing.

Lykes-Duke/Louis Dreyfus Ltd. will provide energy-related services to both wholesale and retail customers in Florida.

Duke/Louis Dreyfus, a leader in major fuel markets, manages price and financial risk as well as being involved in the construction and operation of electric systems. Headquartered in Wilton, Conn., the company provides a broad range of energy products and services to wholesale and retail customers in the United States. Duke/Louis Dreyfus draws much strength from the collaborative strength of Duke Energy Corp., an owner and operator of power-related facilities worldwide, and Louis Dreyfus Electric Power Inc., one of the nation's leading power marketers.

Duke Energy, a subsidiary of Duke Power Co., provides technical experience and financial acumen accumulated over more than 90 years through the successful operation of Duke Power's integrated electric system. Louis Dreyfus Electric Power was the first non-utility in the nation to have a significant volume of wholesale electric business.

Lykes Energy is a Tampa, Florida-based diversified energy products and services company. Its energy subsidiaries include Peoples Gas System Inc., which is Florida's largest and Southeast's sixth-largest natural gas utility, and Peoples Gas Company, the state's largest independent propane supplier with annual sales in excess of 22 million gallons. Another subsidiary, Peoples Cogeneration Co., participates and assists in the development of cogeneration projects in Florida.

EnergyOne

Kansas City, Missouri-based UtiliCorp United is seen by most in both the electric and gas utility businesses as a leader in the area of providing total energy solutions on a nationwide basis through its EnergyOne "brand."

UtiliCorp is a $3.9 billion international electric and gas company with energy customers in 45 states, Canada, Great Britain, New Zealand, Australia and Jamaica. The map in Figure 6-1 illustrates the breadth of the company's business activities and territories.

FIGURE 6-1 UtiliCorp United/Energy One Territory

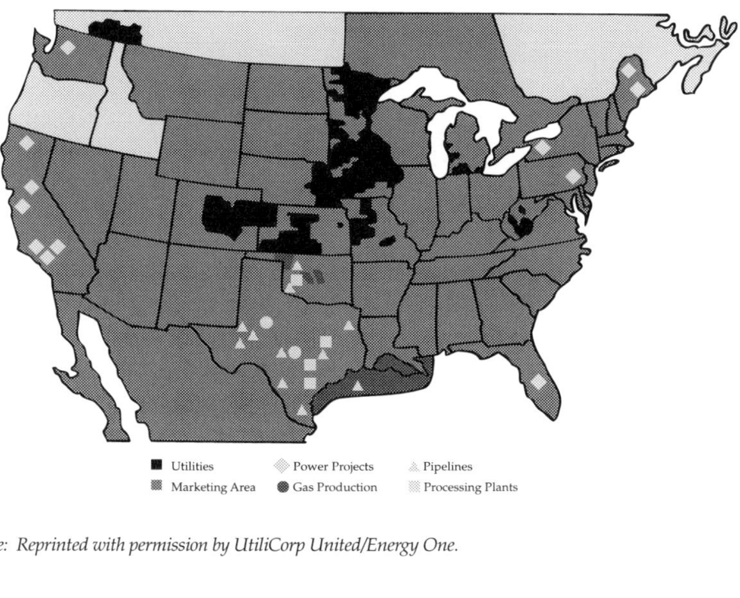

Source: Reprinted with permission by UtiliCorp United/Energy One.

UtiliCorp is a $3.9 billion international electric and gas company with energy customers in 45 states, Canada, Great Britain, New Zealand, Australia and Jamaica.

UtiliCorp United is also engaged in various nonregulated energy- and utility-related businesses in the United States and United Kingdom. Including customers served indirectly, UtiliCorp's businesses reach a total of more than 22 million.

FIG 6-2 GREAT MOMENTS IN ENERGY HISTORY

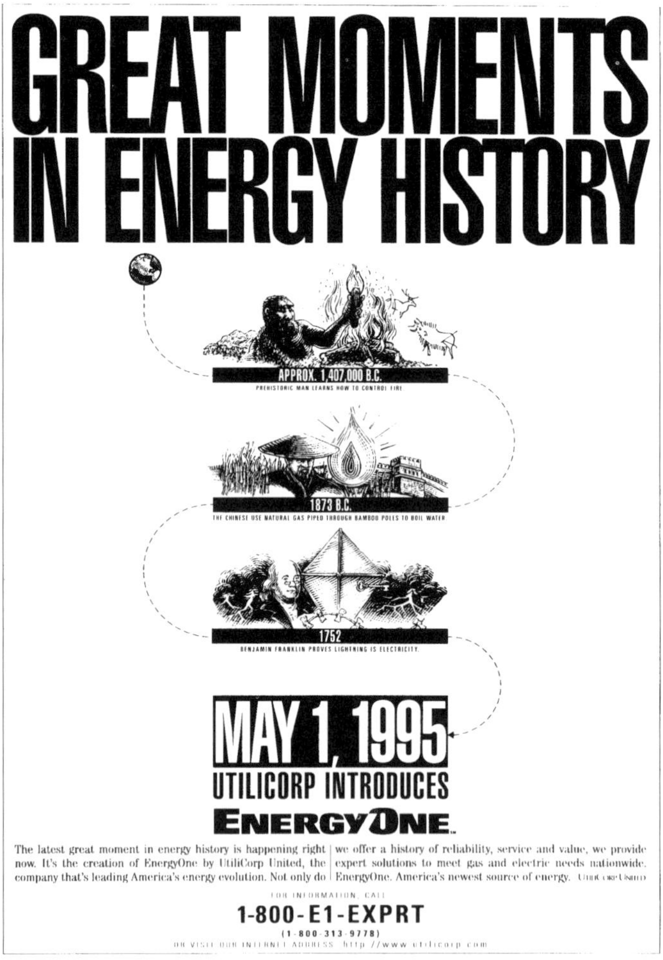

Source: Reprinted with permission by UtiliCorp United/Energy One.

On May 1, 1995, UtiliCorp United launched a national advertising campaign to introduce its EnergyOne brand, which is the first nationally branded line of products and services for electric and gas utility customers. This campaign was launched in the national media through a special advertisement, as shown in Figure 6-2. Of special note is that UtiliCorp United, through EnergyOne, has offered customers a so-called "competitive price guarantee." Refer to Figure 6-3 to learn more about this unique contract.

FIG 6-3 COMPETITIVE PRICE GUARANTEE

UTILICORP UNITED

Competitive
Price Guarantee

UtiliCorp United guarantees
that our EnergyOne services
are competitive with any other company
with the ability to serve you.
If not, we'll apply the difference
in price on our next contract with you.

organization

contract no.

address

city, state, ZIP

contract area

customer signature

company signature

date

Source: Reprinted with permission by UtiliCorp United/Energy One.

EnergyOne offers commercial, industrial, and residential customers a wide variety of *branded* products and services. A sampling of those services is provided in the sidebar shown below.

COMMERCIAL

Term transportation (small-volume customers are guaranteed savings versus regulated sales rates)

Procurement (fee-based service for acquiring gas supplies)

Balancing (fee-based service for providing gas volume balancing on local distribution companies' systems)

Capacity management (securing gas pipeline capacity on a term arrangement for a fee)

Capacity choice (nominated term gas capacity for customers moving third party supply)

Total resource energy evaluation (audit of customersí facilities and recommendations on energy efficiency improvements that qualify for cash rebates)

INDUSTRIAL

Small power services (onsite turnkey generation options)

Gas management services (focuses on unregulated gas services such as transport and capacity release)

Fuel supply services (contract for supply and distribution of primary and secondary fuels)

Special utility services (extends traditional utility service to customer facilities)

Productivity and efficiency services (turnkey tailored technology packages)

Energy profiler services (analyzes the customer's comprehensive energy service)

Environmental services (energy technology solutions to environmental challenges)

Power quality and reliability services (built on "good, better and best" offerings and guarantees to include enhanced supply and customer facility options)

RESIDENTIAL

Appliance financing (low-interest loans for high efficiency equipment)

Appliance repair services (repair contracts at fixed monthly fees)

Appliance sales (full range of appliance products)

Custom power applications (packaged power quality solutions to meet homeowner needs)

Lighting services (new options to meet both security and decorative needs)

Heat pump guarantee (money back offer if dissatisfied with electric heat pump financed and inspected by EnergyOne)

EnergyOne fits all the requirements (and more) of an ESCO, and goes one step further in that it *owns* its own brand of products and services. The owning of products and services and then marketing them to others both inside and outside a utility's service territory is expected to become a trend. UtiliCorp United and its new brand, EnergyOne, are fast becoming models that other utilities can be expected to follow in the future.

EnergyOne is a perfect example of what the Washington International Energy Group terms large *brand* players. As shown in Figure 6-4, these brand players (like EnergyOne) have links to independent power producers and other generators, ESCOs, marketers, large customers, and small niche utilities.

FIGURE 6-4 BRAND PLAYER'S ROLE IN THE NEW INDUSTRY STRUCTURE

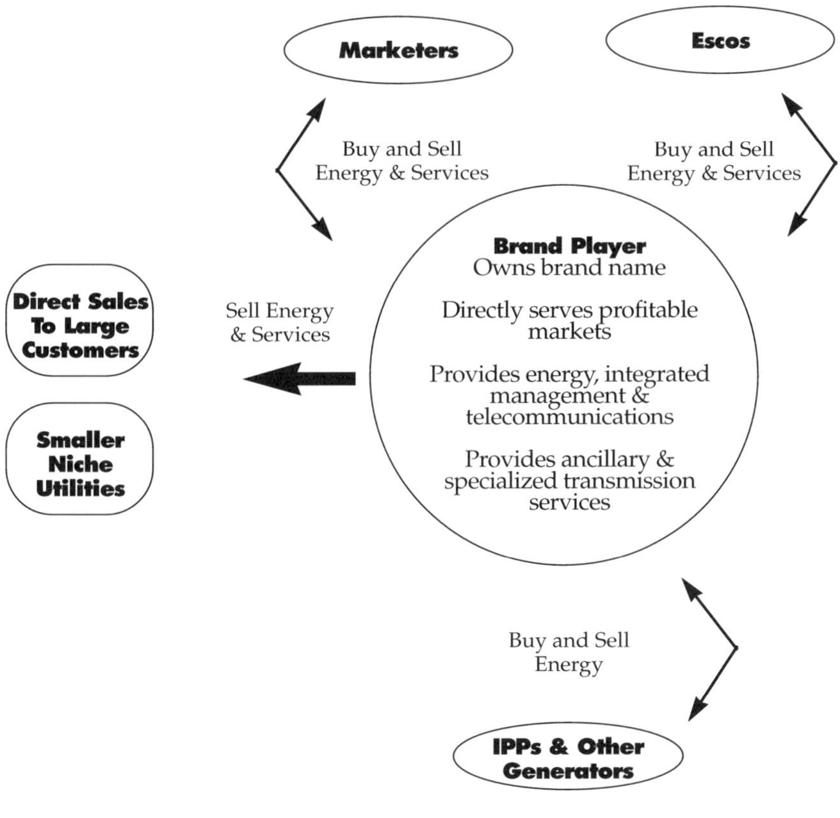

Source: Reprinted with permission by the Washington International Energy Group, Copyright 1996.

PARTNERSHIPS AND MERGERS

As stated in the beginning of this chapter, not only are the formation of ESCOs helping utilities to gain more competitive, and thus customer, advantages, but so too are special partnerships (with other utilities and directly with end-users) and mergers.

In a relatively short period of time, Green Bay-Wisconsin-based Wisconsin Public Service Corp. has struck three "power deals" with other utilities. The latest deal came on May 1, 1996, when the utility, which is a small gas and electric company serving around 400,000 customers, entered into an agreement to supply electricity to the Onconto Electric Cooperative. The long-term agreement calls for Wisconsin Public Service to serve all of the co-op's energy needs by the third year of the contract. The contract guarantees electricity prices to the co-op over the next 10 years and provides for savings in purchased power costs over the life of the contract. Wisconsin Public Service also will provide a variety of products and services to the co-op that will enhance the co-op's economic development programs and assist it in load management to further reduce costs.

Wisconsin Public Service is just one example of a utility teaming up with other utilities through customizing products and services.

This particular contract is one of a total of three the utility has entered into. The two others include a contract with Upper Peninsula Power Co. to provide energy management, after-hours dispatching, and call center products and services along with a signed letter of intent with Manitowoc Public Utilities to provide similar services as well as to operate Manitowoc's 89MW power plant.

Wisconsin Public Service is just one example of a utility teaming up with other utilities through customizing products and services. Another alliance is one recently entered into between CNG Energy Services Corp. and Honeywell, a leader in performance contracting. Under this unique alliance, significant cost savings and flexibility in energy management for operators of hospitals, schools, industrial, and government facilities will be accomplished.

CNG Energy Services is the full-service energy marketing unit of Consolidated Natural Gas Co. It provides electricity, natural gas, other energy forms, and related services to customers across North America. Its wholesale power marketing affiliate in 1995 marketed nearly 2 million MWhr. CNG Energy Services also is a leading natural gas marketer, with managed volumes averaging more than 2.2 billion cubic feet per day in 1995.

According to the companies, the alliance enhances Honeywell's building control programs by making available the expertise and resources of CNG Energy Services in helping to provide the best energy price and supply options to Honeywell customers.

CNG Energy Services initially will offer natural gas supplies and management to Honeywell customers. Electricity supply and management will be offered when deregulation of the power industry permits it.

Still yet another example of an innovative partnering alliance is one struck by Associated Power Services Inc. to supply power management services to the Springfield, Oregon Utility Board.

Associated Power Services Inc. is the power marketing arm of Panhandle Eastern Corp., which operates one of the nation's largest interstate natural gas pipeline networks. Through subsidiaries, such as Associated Power Services, Panhandle markets natural gas and other energy services nationwide as well as in Canada and the United Kingdom.

The alliance between Associated Power Services and the Springfield Utility Board is the first time a power marketer has gained a contract to provide long-term power management services for a municipal utility district.

Under the contract, Associated Power Services will:

1. Manage all of the Springfield Utility Board's system power requirements;
2. Provide daily power scheduling and short-term power sales management; and
3. Provide purchase and storage transactions, contract administration assistance, and energy accounting and billing.

The alliance between Associated Power Services and the Springfield Utility Board is the first time a power marketer has gained a contract to provide long-term power management services for a municipal utility district.

Mergers are yet another way in which energy companies can pool together to offer customers low-cost power and associated value-added products and services.

Probably one of the decade's most interesting cross-energy mergers is that of a merger announced in mid-1996 by Texas Utilities to purchase Enserch Corp., an integrated natural gas company. The purchase price is reportedly $1.7 billion.

As has been demonstrated in this chapter, regardless how alliances and special partnering arrangements are structured, the end goal is to *help* customers realize greater cost savings through enhancements in products and services offered to them. And, as first pointed out in chapter 3, profound strengths are often associated with joining at the hip with other energy companies to provide low cost power and to formulate an integrated menu of customer options.

Chapter 7 will look at the buying and selling of electricity through so-called electronic energy exchanges. It will become evident after reading chapter 7 that strengths gained through the special alliances and partnerships discussed in this chapter will lend much assistance as companies participate in buying and selling electricity.

Buying And Selling Energy Through Electronic Energy Exchange Systems

As presented throughout this book, the new competitive energy marketplace involves new and unique demands. In addition to these new demands, has come a plethora of new and exciting opportunities for energy companies, including being able to effect buy and sell arrangements through using on-line systems.

This chapter will look at how the demand upon energy companies is not only *providing* energy to end-users but also encompasses a new and broader demand—that of procuring and of selling energy—using the best, most cost-efficient methods. This, of course, has always been a paramount concern to energy providers; however, as this chapter will show, this demand has become even more of a concern as companies work to obtain and keep their competitive positions in the new market.

Just leave it to the many breakthroughs in technology to come up with the most interesting (and workable) solutions.

Many opportunities abound in this new technological age. As most other segments of industry in the United States and abroad have discovered, the methodology behind finding new and improved means to effect high efficiency and thus low-cost energy nearly always involves the use of technology, and more specifically *information technology.*

Information is the key word when it comes to the practice that utilities and other (trading) entities—such as power and energy marketers—utilize in purchasing and selling energy. And, as the energy industry has witnessed in the last several years, just leave it to the many breakthroughs in technology to come up with the most interesting (and workable) solutions. Utilities and other entities have used this new technology and added their own bit of creativity to come up with some highly innovative ways to bring about the fastest and most cost-effective methods of purchasing and selling energy.

THE AUTOMATION OF TRADING

While the telephone is still, and will continue to be, an important vehicle in effecting buy/sell arrangements, electronic systems will in the future (if they haven't already) bring far greater efficiencies for utilities and others wishing to trade energy.

While telephone trading has not been totally replaced with electronic trading systems (it may never be *totally* replaced), it would be safe to say that electronic systems certainly will in the future be the so-called "tool of choice" for many. This statement is supported by the fact that Federal Energy Regulatory Commission (FERC) in FERC Order 889 established what is known as the Open Access Same-time Information System rule, which is otherwise known as the OASIS rule. Chapter 2 explains this order in more detail.

Electronic trading systems are by no means "one size fits all" pieces of technology. There are different systems for different circumstances and for different utilities and/or trading entities. By and large, electronic systems may include:

- On-line bulletin board services
- Next-hour energy services
- Bid/ask systems
- Computerized matching services
- Twenty-four hour real-time services

A good example of a 24-hour real-time service is the Electronic Energy Exchange, which was jointly developed by Consolidated Edison Company of New York, Inc. and Cegelec ESCA Corporation of Bellevue, Washington. According to Con Edison, the Electronic Energy Exchange is seen by the utility as a way to expand its trading.

FIGURE 7-1 Cegelec ESCA Help Desk

Cegelec ESCA is the sole marketer and operator of the Electronic Energy Exchange. As illustrated in Figure 7-1, the company has set up an operations center and help desk at a high-level security site in Bellevue, Wash.

According to Con Edison, the Electronic Energy Exchange is seen by the utility as a way to expand its trading.

Source: Permission to reproduce granted by Cegelec ESCA Corporation.

FIGURE 7-2 Electronic Energy Posting Board Screen Capture

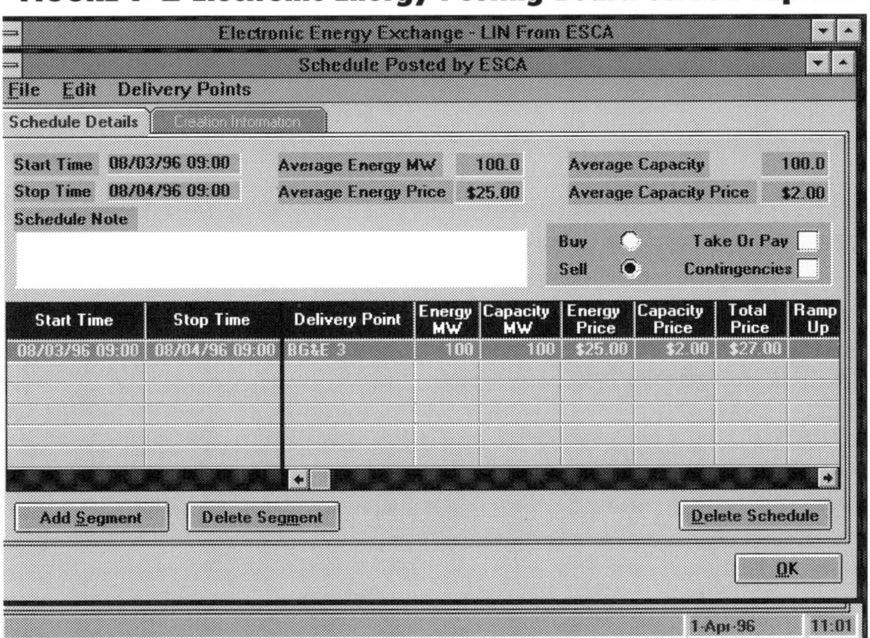

Source: Permission to reproduce granted by Cegelec ESCA Corporation.

FIGURE 7-3 Electronic Energy Exchange Schedule Creation Screen Capture

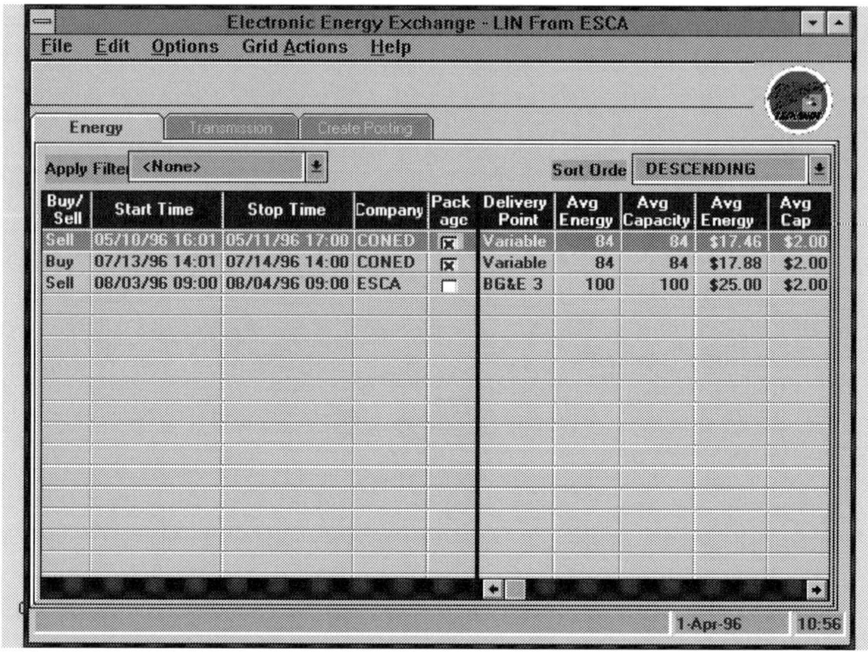

Source: Permission to reproduce granted by Cegelec ESCA Corporation.

The Electronic Energy Exchange, which is designed especially for electricity trades, combines in its operation electronic bulletin boards or EBBs and the use of private telephone negotiations. Open seven days a week, 24-hours a day, the system allows traders to negotiate multiple transactions simultaneously in real time. Refer to Figure 7-2 and Figure 7-3 for a look at two of the system's screens: the posting board screen and the schedule creation screen. In order to utilize the exchange, a subscriber would pay a flat monthly fee to post any number of offers at any time for any duration.

So far, according to Con Edison and Cegelec ESCA, subscribers are those who want more than bulletin board postings and more flexibility than next-hour trading can offer. A typical subscriber would be one that conducts high-volume trading, is seeking to improve its profit margin through negotiated deals, and is wanting access to the market for more than just a few hours.

Subscribers have the benefit of receiving all software, hardware, and communications equipment from the companies in order to take part in the service. Cegelec ESCA also installs the Pentium PCs, all necessary software, and the Alcatel frame relay network connection. It is a Microsoft Windows-based application, so most people in the industry would probably find this particular service user-friendly.

NEXT-HOUR ELECTRONIC TRADING SYSTEMS

Next-hour electronic trading systems have their own set of highly valuable characteristics. This section will focus on two next-hour electronic energy exchanges: CPEXTM and AIMS (Automated Interchange Matching Systems Inc.). It also will provide some details on an announcement in California about a new exchange, called WEPEX, that will form both a power exchange and an Independent System Operator or ISO.

CPEX

Known in the energy industry as the nation's first competitive next-hour electronic energy exchange, energy currently is being offered, sold, scheduled, and delivered through the use of CPEX.

The unique next-hour energy trading system, which is owned and operated by Continental Power Exchange Inc. of Atlanta, Georgia, went on-line May 15, 1995. Much like other on-line systems, membership is required in order to participate in the next-hour system.

Continental integrated a transmission EBB (electronic bulletin board) with CPEX about a month after the system went on-line. The integration, according to the company, provides "one-stop shopping" for transmission users and energy traders.

CPEX currently operates in four North American Electric Reliability Council (NERC) regions in the Eastern Interconnection: Mid-America Interconnected Network (MAIN); Mid-Continent Area Power Pool (MAPP); Southwest Power Pool (SPP); and East Central Area Reliability Coordination Agreement (ECAR). Refer to chapter 3 for a map of NERC regions.

Known in the energy industry as the nation's first competitive next-hour electronic energy exchange, energy is currently being offered, sold, scheduled and delivered through the use of CPEX.

CPEX is working with a group of Western utilities to begin real-time energy trading in the near future and hopes to expand the system throughout the country. Continental reports overwhelming support and use of the system as more than 40 companies are either currently participating in the system or are beginning the process.

The users of the CPEX system are varied. Currently, they include the following groups:

- Investor-owned electric utilities
- Municipal utilities
- Rural electric cooperatives
- State districts
- Federal power marketing agencies (two at the writing of this book were participating)

The logistics of the CPEX next-hour trading system enables participants to trade energy on a comparable basis through which all parties are able to simultaneously learn of and act on each others' buy and sell offers.

Perhaps it is the design of CPEX itself that lends the system its uniqueness and attractiveness to users. A wide-area telecommunications network is coordinated by a high-speed computer system that is capable of processing thousands of energy transactions each hour.

Automatically identified for the participant is the least-cost feasible transmission path for all energy offers. So, CPEX actually will search out where transmission is available, provide the cost of the transmission, and then arrange the most economical delivery path for the transaction.

The buying and selling of energy through CPEX is a point-and-click computer operation. Sales of energy are effected on-line. The system will arrange the delivery path and will provide dispatchers with a schedule prior to implementation. With CPEX, all parties to a particular transaction are aware of each others' identities and transactions are able to be curtailed prior to or during the delivery hour.

CPEX employs a 24-hour help desk that participants can access via a toll-free number for assistance when they need it.

The CPEX system has guarded itself, however, against some potential problems that could arise. For instance, the system is set up with redundant telecommunications and computer systems so that no single point of failure exists. In addition, an uninterruptible power supply with backup generators ensures continuous operation.

AIMS

At the writing of this book, AIMS or Automated Interchange Matching System Inc. appears to be the only next-hour energy trading system other than CPEX that is actually up and running.

AIMS was formed in 1994 and had its membership finalized in May 1995. The system began next-hour trading on March 18, 1996 and operates on a seven-day-a-week, 24-hour-per-day schedule.

As of Dec. 31, 1995, approximately 30 entities had joined AIMS. Those able to gain membership must be electric utilities, power marketers, or independent power producers. In addition, entities that meet the following criteria are eligible: one that is subject to, or eligible to apply for, an order under Section 211 of the Federal Power Act or operates a control area or agrees to arrange for control area services to be performed for it by another utility in connection with the AIMS matching system.

AIMS is unique in that it is a non-profit corporation which was formed by 26 investor-owned federal, municipal, and rural electric cooperatives that operate in the eastern half of the United States.

Members of AIMS provide information so that the matching system can compile a matrix of terms, conditions, transmission requirements, and other data that applies to a potential transaction between members.

AIMS is unique in that it is a non-profit corporation that was formed by 26 investor-owned federal, municipal and rural electric cooperatives that operate in the eastern half of the United States.

Once this basic information has been entered into the system database, buy or sell quotes may then be entered by the members. Quotes are either cost- or market-based. Members participating in AIMS are entering into bilateral contracts for the purchase and sale of non-firm energy. Members essentially only deal with and negotiate with other members. AIMS is not a participant in the process.

Once members post buy or sell price quotes during one hour for electric energy to be generated and delivered during the next hour, the AIMS system matches the highest buy quotes with the lowest sell quotes. Once this is accomplished, the system then transmits the results electronically to the matching members in time to schedule and complete the transactions.

AIMS' central computer and other communications equipment is on site at Tennessee Valley Authority facilities in Chattanooga,

Tennessee. Members are linked to the network via their own computer systems which are connected to the AIMS host computer by high-speed data communications links.

WEPEX

On April 29, 1996, Southern California Edison, San Diego Gas & Electric, and Pacific Gas & Electric-California's three investor-owned electric utilities (IOUs)—filed with the Federal Energy Regulatory Commission and the state public utilities commission joint recommendations for the operations of a power exchange (WEPEX) and independent system operator (ISO).

WEPEX's steering committee includes advisory representatives from large and small customers, independent power producers, power marketers, municipal utilities, and the California Energy Commission.

According to the three IOUs, through the exchange, generators would compete against each other to offer the lowest cost power to consumers. The exchange would post the most competitive spot market price for generation in an open market for everyone to see. In this way, it operates much like a stock exchange.

Through the ISO, every generator would be assured access to the state's electricity transmission system. The ISO will take control of the power grid to coordinate the generation and transmission of electricity and ensure a reliable, safe power supply. Both the exchange and the ISO will be independent of California's utilities and will regulated by FERC.

See the sidebar, "Independent System Operator Role and Responsibilities," for specifics about the ISO.

INDEPENDENT SYSTEM OPERATOR ROLE AND RESPONSIBILITIES

- Coordinates the day-ahead scheduling and real-time balancing for all users of the grid.
- Complies with all NERC and Western Systems Coordinating Council (WSCC) operating and reliability standards.
- Controls the dispatch of the combined transmission facilities of the participating utilities.
- Provides nondiscriminatory open access to the transmission grid and ancillary services to all users.

- Manages transmission congestion and constraints on network basis. All users would be subject to the same access protocols and prices.

- Procures ancillary services on a competitive and unbundled basis and offers them to users initially as bundled services.

- Performs its settlement function.

- Provides transparent information flow.

The power exchange has five major responsibilities. They are:

1. To provide a competitive spot market for power with published hourly prices.

2. To allow power producers to compete using nondiscriminatory and transparent rules for bidding into the exchange.

3. To take supply bids from generators and demand bids for power from utilities, retailers, and power marketers, etc.

4. To rank bids and submit to the ISO a preferred least-cost dispatch schedule for delivery of power.

5. To ensure a visible market clearing price that will allow customers to make efficient purchasing decisions and adjust their consumption.

The power exchange's 17-member governing board would be made up of three non-utility generators, four end-users, two non-stakeholders/public; three buyers/sellers; and five distribution companies. It would take 12 votes to adopt most measures and only six to veto most measures.

The power exchange would collect supply and demand bids one day ahead of actual production or use from generators and retailers. The aggregate match between supply and demand would determine hourly energy prices. Generators would be permitted to specify the following three components of their bid: (1) Energy cost, (2) No-load cost, and (3) Start-up cost.

Customers would be allowed to specify the maximum price level they would be willing to pay. For example: 100MW if the price is four cents or less or 80MW if the price is more than four cents. Powers exchange bids are evaluated for a 24-hour period. It is hoped by the three IOUs that start-up of the ISO and the power exchange would be in place by Jan. 1, 1998.

CHAPTER 8

Managing Risk In A Competitive Environment

It is fitting that this book end with a chapter on how utilities as well as other players can manage risk in the face of unprecedented market changes. The chapters in this book have in some way made either direct or indirect inferences to deregulation and competition. This new market-based structure of the power industry, after all, is the primary, underlying driver of the practice known as energy marketing.

Deregulation and competition in the electric utility industry has spurred an onslaught of activity among players other than utilities that have entered or stepped up their involvement as being *energy service providers.* These other competitive players include independent power producers, industrial cogenerators, power (and energy) marketers, and natural gas marketing companies.

Utilities are finding they now have competitors that must be dealt with. As the old adage goes, "Lead, follow or get out the way." Utilities are in no mood to get out of the way or follow. So, in their quest to lead, many have formed their own subsidiaries (sometimes internal, sometimes external) through the creation of energy or power marketing firms, that will assist them in finding new wholesale customers outside their current areas of service.

Whether created internally through a department or created externally through the formal creation of a power or energy marketing firm, utilities have found a *safe haven* wherein they can place a team of experts on the risk management track.

In the future, as deregulation takes hold and spreads out to other classes of customers, such as at the residential level, utilities will be utilizing their subsidiary arms or departments (if internally operated).

These will help manage the risk associated with seeking out new provider opportunities as well.

This increasingly competitive operating environment has essentially forced the hand of many utilities to take immediate action on retaining their market position. Utilities have found, however, that they must play their hands as systematically as feasible. Risk management is the key.

RISK COMES WITH THE TERRITORY

Risk is the net effect of a competitive environment. With the many changes that competition is bringing about, utilities must shoulder a substantial amount of risk as they cope with the realities of change.

Through operating in a competitive environment,
utilities will need to develop ways to manage
risk in two different areas:
power sales and power purchases.

Most utilities are finding that their risk management strategies are becoming as integral to their day-to-day operations as are their basic business and operations strategies.

Because risk is such an integral part of the current and future utility environment, it becomes essential that utilities learn how to manage and thereby reduce their risk as much as possible.

Profitability in the age of competition is at the heart of why so many utilities are, or will in the future, either have internal risk management tools or external risk management tools. These will be at their disposal to help bolster, retain, and maintain their competitive advantage.

A NEW BUSINESS STRATEGY NEEDED

Through operating in a competitive environment, utilities will need to develop ways to manage risk in two different areas: power sales and power purchases.

In an environment where traditional utility customers are able to choose their own providers, utilities must work to engender a new business (and thus risk) strategy is dealing with power sales. Being the supplier of choice means that utilities must recognize what their strengths and weaknesses are and develop strategic plans and resources to back important decisions about their supply arrangements. It is here that risk management strategies can be of great benefit.

Volatile prices, greater market risks, and a fluctuating demand will mean for most utilities that more attention must be paid to managing their energy purchases. As discussed in chapter 1, electricity is now considered a commodity. As such, utilities can employ many different types of risk management for both their sales and purchasing procedures. From forming an internal or external power or energy marketing entity to utilizing financial market risk strategies (such as those available through futures), utilities can, and are in many cases, already taking advantage of the many avenues open to them to help assess and deal with the multitude of risks presented by competition.

As this chapter will illustrate, there are many tools which utilities can use to help manage risk in a competitive atmosphere. This chapter in no way lists all the ways which utilities can employ the use of risk management strategies, but does give some of the more popular tools currently being utilized.

THE ROLE OF MARKETERS AND BROKERS

It is commonplace to find electric utilities getting into the business of marketing energy. Several utilities have actually formed power marketing or energy marketing subsidiaries who have this function as their primary purpose of existence.

As stated elsewhere in this book, natural gas companies as well as independent entities are also forming energy marketing companies. Appendix D provides a listing of power and energy marketers and contact names. Appendix D lends testament to the fact that there are many different interests in the business of marketing. Also of interest is Appendix E, which is a detailed power marketer business plan.

Chapter 1 fully explains the differences between a *power marketer* and an *energy marketer*. Refer to that chapter for a full description of these terms. For a very brief recap, however, suffice it to say that a power marketer is a marketer of electricity only, while an energy marketer is a marketer of various forms of energy, which includes electric power, natural gas, cogeneration, and other energy alternatives.

This chapter will look at the marketer's function, and will apply to marketing in general whether the entity markets only electricity or various forms of energy.

There are two different types of marketers: affiliated and independent. Under Federal Energy Regulatory Commission guidelines, affiliated marketers are comprised of any group that has more than a five-percent or 10-percent ownership by an electric utility, qualified facility, or independent power producer. Currently, utilities comprise 10 percent of the affiliated marketer segment. Independent marketers are just that—and fall outside the parameters listed above for affiliated marketers. All types of marketers, both affiliated and independent, are subject to FERC reporting requirements, as detailed above.

It must be noted that brokers also carry a significant amount of weight in arranging bulk power transactions. While brokers will not receive as much attention in this chapter as the marketers, their role should not be downplayed.

There are at least two main differences between a marketer and a broker. First, brokers do not assume financial risk. The role of the broker is to bring together a willing seller and a willing buyer without actually taking title to the power. Marketers actually take title to the power, thus their attractiveness in reducing a utility's financial risk. Secondly, brokers differ from marketers in that they currently are not regulated by FERC.

As shown in Appendix D, the ranks of marketers are growing at a fast pace. The rise in the number of marketers and of their transactions at the wholesale level are largely due to their unique ability to offer services that utilities traditionally have not developed. Marketers are able to unbundle and then rebundle differentiated products to meet the needs of a vastly changing market.

Through this unbundling and rebundling activity, marketers are able to provide high-impact and high-demand services. In the physical market, marketers are able to provide the following services: system firming, shaping, dispatchability, displaceability, and storage. In the financial area, marketers are able to offer the following: flexible contracts, flexible pricing, and various financial instruments, such as futures. Futures are discussed later in this chapter.

These attributes particular to marketers make them highly attractive to utilities that generate electricity and are looking for new, innovative ways to market the power they produce. As stated earlier, the marketing of power outside a utility's traditional service area is tantamount in clinching success in a competitive environment.

Utilities that once found themselves having to enter into long-term power purchase contracts are now finding that they can go with short-term contracts when they employ the use of marketers. This implies two things: that the contracts are viewed by utilities as *safer* and second that utilities like the idea of sharing or relinquishing altogether their financial risk through shifting it to the marketer.

FINANCIAL MARKET ADVANCEMENTS: NYMEX

One of the most exciting and talked about occurrences in 1996 is the New York Mercantile Exchange's (NYMEX) launching of both its

Palo Verde and California-Oregon Border (COB) electricity futures contracts. The sidebar below lists the terms of the two contracts. The cover photo on this book depicts the trading floor.

PALO VERDE AND COB INDEX TERMS

Contract Unit: 736MWhr delivered over a monthly period.

Listed Months: 18 consecutive months.

Delivery Rate: 2MW throughout every hour of the delivery period (this can be amended upon mutual agreement of the buyer and seller.)

Delivery Period: 16 on-peak hours—hour ending 0700 prevailing time to hour ending 2200 prevailing time. (This can also be amended at the time of delivery by mutual consent of the buyer and seller.)

Termination of Trading: Trading in the delivery month shall cease on the fourth business day prior to the first day of the delivery month.

Prices and Fluctuations: Prices shall be quoted in dollars and cents per MWhr. The minimum price fluctuation will be $.01 per MWhr. The maximum daily price fluctuation will be $3 per MWhr above or below the preceding day's settlement price (the basic maximum fluctuation). Expanded limits will apply when the contract settles at the maximum limit.

Scheduling: Buyer and seller must follow Western Systems Coordinating Council scheduling practices.

Launched March 29, 1996, the contracts were offered for electronic trading on NYMEX ACCESS℠ on April 1 as part of the April 2 trading session.

Electricity not only is gaining commodity recognition at NYMEX, the largest physical commodity futures exchange in the world, but also is gaining this recognition by regional exchanges, such as the Boston Stock Exchange, as shown in Figure 8-1.

FIGURE 8-1 Boston Stock Exchange

Source: Permission to reproduce granted by Boston Edison Co.

WHY USE FUTURES?

According to NYMEX, the following are commercial applications of using energy futures:

1. **Independent power producers** can sell futures contracts to lock in a sales price of their power.

2. **Large buyers** can purchase futures to cap their purchase price.

3. **Power marketers**, who have exposure on both sides of the market, can hedge with futures to mitigate that risk, buying or selling contracts as appropriate.

4. **Integrated utilities** may be able to use futures to protect revenue targets. As performance-based rate making gains acceptance, futures can be used to achieve performance benchmarks.

Futures markets specifically help businesses manage their price risk, according to NYMEX, by providing a means of hedging and of matching buyers and sellers of a commodity with parties who are either more able or more willing to bear the risk or who have inverse risk profiles. Two good examples would be an independent power producer who sells an electricity futures contract to effectively place a floor under his sales price while an industrial consumer might buy electricity futures to cap his acquisition cost.

NYMEX is uniquely positioned to offer to the electricity industry what it has offered over the years to other physical commodities. The exchange launched its heating fuel oil futures contract in 1978, which was the first successful energy futures contract. The contract was introduced during a period of extreme volatility in the oil markets, and has evolved into an instrument used to hedge price risk in a broad range of refined products.

NYMEX also introduced futures and options contracts for crude oil, the most actively traded contract on a physical commodity; natural gas; gasoline; and propane futures.

CONCLUSION

Risk management is by no means a new concept to utilities or to other players involved in the marketing of energy. The goal of this chapter (and of the book) has been to show that while there are many new activities currently being undertaken by energy companies, there are still some traditional activities that simply are being *beefed up* in order to meet the new competitive challenges of the future.

From the commingling of the electric utility industry with the natural gas industry and vice-versa (chapter 3) to the competitive implications of retail wheeling (chapter 4) and wholesale wheeling (chapter 5) to the formation of special marketing relationships (chapter 6), effecting buy and sell arrangements electronically (chapter 7) and finding new and innovative ways to handle market risk (chapter 8), this book has attempted to cover a wide array of topics associated with energy marketing couched against the deregulation and competition aura that now surrounds the power industry. While by no means complete, the attempt at covering these areas of concern has been enlightening.

Consider the following observation by industry expert Joseph P. Kearney, president and CEO of U.S. Generating Company (USGen), a national power supplier with interests in electricity generation, power marketing, natural gas transmission and storage, and fuels management.

Kearney contends that today's independent power producer (IPP) will soon be a distant memory. What is going to happen, he says, is that IPPs will be replaced with what he calls the "Competitive Power Supplier."

Kearney says that restructuring in the power industry is quickly blurring the functional distinctions between utilities and IPPs. He said all market participants of the future must adapt, adapt, adapt in order to survive the competitive environment. Participants must:

- Adapt to market-based commodity pricing,
- Adapt to short-term contracts,
- Adapt to the disaggregation of generation, transmission, and distribution.

This Competitive Power Supplier, Kearney says, will be a hybrid of the IPP and power marketer and the generation and bulk power components of utilities. As stated in chapter 6, price is king and a major driver in the push toward a Competitive Power Supplier.

Other drivers of this new power supplier, according to Kearney, include:

- new entrants to the power supply market
- the introduction of two electricity futures contracts by NYMEX
- growing price transparency in the bulk power trading market
- the enormous growth of power marketers
- the pressure to resolve difficult transition issues
- the impact of mergers and acquisitions on market dynamics.

USGen, which currently is an IPP, is in the process of repositioning itself from a generator to a diversified energy company that combines a multitude of disciplines and services including generation, power marketing, fuels management and trading, asset acquisition and management, and merchant power plant development.

GLOSSARY

A

Access Charge - A term from the telephone industry. Also called a transition charge, it recovers costs required of utilities while under regulation.

Affiliated Power Marketer - Commonly known as APMs, these are affiliates of established energy industry companies that seek and achieve power marketer status with the Federal Energy Regulatory Commission (FERC). Under FERC guidelines, an APM is any group with more than five-percent or 10-percent ownership by an electric utility, qualified facility, or independent power producer.

Affiliate Transactions - Transactions between companies, one of which owns a controlling interest of the other or all of which are controlled by the same holding (or parent) company.

Aggregators - Marketing companies that pool energy from various sources into packages for resale to end-users.

Allowance For Funds Used During Construction (AFUDC) - Refers to a utility's accrual of return on its invested capital during the construction period, which is then added to the direct cost of the facility for the purpose of determining the rate base on which it is allowed to earn a return after the facility goes into operation.

Ancillary Services - Products that must be delivered to the interconnected electric utility system to ensure it will operate properly. Ancillary services from generating resources include: voltage control, reactive power dispatch, spinning reserve, quick-start reserve and load following.

Antitrust - Laws or policies that encourage competition by curtailing monopolistic power and unfair business practices.
Appreciation - The increase in the value of property.
Arbitrage - Buying securities in one market and selling them at the same time in another to make a profit on the price discrepancy.
Arms Length Transactions - When dealings transpire between unaffiliated companies, they are known as arms length transactions.
Asset - Items that are readily converted into cash.
Automated Interchange Matching System, Inc. (AIMS) - A non-profit corporation that was formed by 26 investor-owned federal, municipal, and rural electric cooperative utilities. The corporation has created a computer-based network that automatically matches quotes for the purchase and sale of one-hour blocks of non-firm electric energy to produce greater benefits for members. Membership is available by calling Robert E. Tyler, Commonwealth Edison Co. bulk power marketing manager, at (312) 394-3409. AIMS was formed in 1994 and began next-hour trading on March 18, 1996. The system is operational 24 hours, 7 days-a-week. Current members follow: Alabama Electric Co-op, Alabama Municipal Electric Authority, Arkansas Electric Co-op, Cinergy, City of Tallahassee, Commonwealth Edison, Duke Power, East Kentucky Power, Enron Power Marketing, Entergy Power Inc., Florida Power, Gainesville Regional Utilities, Jacksonville Electric Authority, Kentucky Utilities, Louisville Gas & Electric, LG&E Power, Municipal Electric Authority of Georgia, NC Municipal Agency No. 1, Oglethorpe Power, PECO Energy, Piedmont Municipal Power Agency, Santee Cooper, South Carolina Electric & Gas, Southern Company Services, Tennessee Valley Authority, Union Electric Co., Virginia Power.
Avoided Cost - The Federal Energy Regulatory Commission (FERC) defines avoided costs as the incremental cost to an electric utility of electric power that the utility would generate itself or purchase from another source if it did not purchase that power from Qualifying Facilities or QFs. FERC regulations implement the Public Utility Regulatory Policies Act (PURPA), which requires utilities to purchase electricity from QFs at a price at or below their avoided costs.

B

Balance Sheet - This is an exhaustive listing of all assets, liabilities and net worth that show the financial condition of a company at any given time.

Balance Sheet Financing - A conventional method of financing capital projects where a company solicits debt on the basis of its overall financial strength and commits its companywide assets as security.

Balancing Item - Represents the difference between the sum of the components of natural gas supply and the sum of the components of natural gas disposition. These differences may be due to quantities lost or to the effects of data reporting problems.

Base (Cushion) Gas - Base Gas is the volume of gas needed as a permanent inventory to maintain adequate underground storage reservoir pressures and deliverability rates throughout the withdrawal season. All native gas is included in the base gas volume.

Base Load - Refers to the lowest load level during a utility's daily or annual operating cycle. Base load, then, refers to supplies of natural gas or electricity intended to serve customer demand year-round.

Basis - A geographic price differential between a particular market and the delivery point as specified in an exchange-traded commodity contract.

Basis Differentials - Refers to the price differences related to timing, quality, sales class, and location features of commoditiy market indexes.

Basis Point - The movement of interest rates or yields expressed in hundreds of a percent.

Basis Risk - Basis risk describes the risk that the value of one instrument may not move up or down in equal proportion with the value of another instrument—usually a hedge.

Bear Market - A period marked by declining stock prices.

Bearer Bond - A bond where the owner's name is not registered on the books of the issuing company. Interest and principal is paid to the bond holder.

Bearer Stock - Stock certficates that are not registered under any name. They are negotiable without endorsement.

Big Board - A common second reference to the New York Stock Exchange.

Biomass - Any organic material that can be used as a fuel to generate energy. Examples of biomass fuel include agricultural wastes, yard and lawn wastes, and animal waste. All biomass fuel matter can be easily converted to energy-producing fuels using available technologies.

Blanket Certificate - Refers to the generic regulatory authorization to purchase, transport, or sell natural gas or electricity under specific, agreed-upon terms, but without the specification of particular suppliers or customers.

Blue Chip Stock - Company stock that is characterized by its long-established record of making money and paying dividends.

Blue Book Restructuring Model - On April 24, 1994, the California Public Utilities Commission (CPUC) issued a 64-page proposed order to restructure the state's electric utility industry. The proposal is informally referred to as the "Blue Book," and formally known as the "Order Instituting Rulemaking and Order Instituting Investigation Governing Restructuring California's Electric Services Industry and Reforming Regulation." The CPUC's objective in issuing the order was to lower the cost of electric service to the state's residential and business consumers without sacrificing electric utilities' financial integrity. Consumer choice through direct access or retail wheeling was a cornerstone of the CPUC's proposal. The proposal called for a phase-in of direct access between the years 1996 and 1997 for industrial customers, 1999 for commercial customers, and the year 2002 for residential customers. The CPUC has held many meetings regarding its proposal. It has accepted comments on the proposal from all affected stakeholders (including electric utilities, natural gas utilities, end-users, environmental groups, trade groups, and many others). The intent to get stakeholders' viewpoints on the proposal culminated in a Dec. 20, 1995 landmark restructuring decision made by the CPUC in which it voted by a 3-to-2 margin to adopt a plan that would create a competitive electric utility market structure. The adopted plan provides for competition and consumer choice beginning in 1998 and would become the nation's first statewide deregulation program. Under the plan, by Jan. 1, 1998, a representative number of customers from all customer groups will be allowed to participate in the first phase of direct access, which would last for one year. After the first year, there would be no limitations on participation in direct-access service except for technical constraints. The CPUC's vision is to have all customers actively participating in direct access by the year 2003. More detailed information on the plan, which has not yet been finalized, can be obtained through: http://www.cpuc.ca.gov or through gopher.cpuc.ca.gov or through ftp.cpuc.ca.gov on the Internet.

Boiler Fuels - These are fuels that are suitable for the generation of steam in large industrial or electric utility boilers. Boiler fuels most often are the lowest-cost fuel per unit of heating value under the specific circumstances. Natural gas, residual oil, coal, and uranium are the dominant boiler fuels.

Bond Ratings - Grades given by credit-rating agencies to corporate and municipal debt securities, based on the borrower's expected ability to repay. The higher the grade, the lower the interest rate. Two major ones are Moody's Investors Service Inc. and Standard & Poor's.

Bond - A certificate issued by a corporation or government stating the loan amount, the interest to be paid, the time for repayment and the collateral pledged if payment cannot be made. Repayment generally is long-term—usually seven years or more.

Book Value - The difference between a company's assets and liabilities. The book value per share of common stock is computed by dividing the book value by the number of outstanding common shares.

British Thermal Unit (Btu) - Commonly referred to as Btu, a British Thermal Unit is the heat required to raise the temperature of one pound of water by one degree Fahrenheit at or near 39.2 degrees Fahrenheit.

Bull Market - A period of increasing market prices.

Bundled Resales - Refers to the sale of natural gas at wholesale or to end-users at a single price that incorporates all transport and storage costs as well as the first-sale price of the gas commodity.

Burner Tip - Signifies the delivery of energy to the end-user.

Buy-Sell Transactions - Consists of the exchanges of natural gas where a local distribution company or LDC or marketer that holds firm capacity on interstate pipelines essentially sublets that capacity in the secondary market.

Bypass - Direct sales by producers, pipelines or marketers to end-users.

C

California-Oregon Border Electricity Price Index - Also known as the COB Index, the California-Oregon Border Electricity Price Index is first in a series of market-based electricity price indexes published on the West Coast since the second half of 1995 by Dow Jones. It is a collaborative effort of a group of investor-owned electric utilities, power marketers, municipal utilities, and federal power agencies that buys and sells electric energy at the California-Oregon border.

Call Option - The right to buy a commodity, whether futures or physicals, at a specified price by a certain date. *Compare put option.*

Cap - In order to hedge or protect itself against unforseen price increases in excess of the market's ability to pay, for example, in purchasing natural gas, an electric utility may purchase an option that sets a so-called "ceiling" price for a specified amount of time. In doing this, the utility retains the right to benefit fully from natural gas prices below the stated maximum price. This hedging strategy, which is often used, is termed a cap. *Compare collar.*

Capacity - Refers to the ability of a power plant to produce a given output of electric energy at an instant in time. Electric capacity is measured in kilowatts or megawatts.

Capacity Reservation Tariff (CRT) - Proposed in a notice of proposed rulemaking (NOPR) April 24, 1996 by the Federal Energy Regulatory Commission. The CRT NOPR (formally the Capacity Reservation Open Access Transmission Tariffs NOPR) was issued in conjunction with the adoption of two historic electric utility industry restructuring rules—FERC Order 888 and FERC Order 889. Under the CRT, utilities and all other power market participants would reserve firm rights to transfer power between designated receipt and delivery points.

Capital - Money, equipment or property used in a business by a person or corporation.

Capital Gain, Capital Cost - The difference between what a capital asset costs and the price it brought when sold.

Capital Structure - Refers to a company's invested capital. In specific terms, capital structure is the relative amounts of various categories of debt and equity in the company's total capital.

Certificate - Known as a "certificate of public convenience and necessity," a certificate is a permit or license that is granted by a regulatory commission to a public utility that allows the utility to provide a specific service or to build and/or operate a certain facility.

Charge Off - A loan that is no longer expected to be repaid and is written off as a bad debt.

Chicago Board of Trade - Better known as "CBOT," the largest commodity trading market in the United States.

Chicago Board of Options Exchange - An exchange set up by the Chicago Board of Trade.

City-Gate - A point or measuring station at which a gas distribution company receives gas from a pipline company or transmission system.

Cogeneration - The sequential production of electricity and thermal energy from a single-fuel source. The resultant thermal energy is most often used in an industrial processing or heating application.

Collateral - Stock or other property a borrower must relinquish if unable to repay a loan.

Collar - A collar is most often used as a hedging strategy and is designed to reduce risk in the commodity markets. A collar essentially guarantees a specified range of gas prices, for example, for an electric utility, over a specified period of time. In some ways a collar is similar to a cap, however a collar permits the utility to specify a maximum amount that it will pay for the gas and without the cost of an up-front premium for price protection. In essence, a collar enables the utility to pay for price protection by relinquishing some portion of the benefit that the utility might get if the market price of natural gas should happen to fall below an agreed-upon strike price. *Compare cap.*

Combined Cycle - Refers to combined-cycle generation, which is a high efficiency power production process. In a typical combined-cycle power plant, combustion turbines burn natural gas or other fuels to generate electricity in the first cycle. In the second cycle, the exhaust heat is captured, rather than vented into the atmosphere and is used to generate steam which drives steam turbines to supply additional electric power. By using heat that otherwise would be wasted to generate additional power, the combined-cycle unit produces cost savings as well as increased operations efficiency. Lower overall plant emissions also come as a result due to greater efficiency, which means that more power is being produced per unit of fuel.

Combined Utility - A public utility that is either privately owned or owned by a municipality that sells both natural gas and electricity.

Commercial Consumption - Gas used by non-manufacturing organizations such as hotels, restaurants, retail stores, laundries, and other service enterprises, and gas used by local, state, and federal agencies engaged in non-manufacturing activities.

Commercial Customers - Commercial customers include the category of energy used by the retail and wholesale trade, service establishments, hotels, offices and public institutions.

Commercial Paper - A short-term negotiable instrument whereby industrial or finance companies obtain cash after agreeing to pay a specified amount of money on the due date.

Commodity Charge - A customer charge for utility service that is proportional to the amount of natural gas or electricity the customer actually purchased.

Commodities Futures Contract - A contract to buy or sell a specific amount of a commodity at a specified future date.

Commodities Futures Trading Commission (CFTC) - A U.S. government agency that oversees the activities of the United States' 11 futures exchanges. *See futures exchanges.*

Comparability - The condition that exists when all parties in an energy trading scenario have equal access to the marketplace and can see and take advantage of each other's offers to buy and sell energy.

Competitive Bidding - Refers to a process that energy companies utilize in many states to select suppliers of new capacity and energy. For example, under competitive bidding, an electric utility solicits bids from prospective power generators to meet current or future power demands. When offers from independent power producers began exceeding utility needs in the mid-1980s, utilities and state regulators began using competitive bidding systems to select more fairly among numerous supply alternatives.

Constant Dollars - In relationship to past or future costs or prices, constant dollars refer to figures that are quoted in inflation-adjusted terms relative to a certain year.

Construction Work in Progress (CWIP) - Utility property that regulators allow to be placed in the rate base even though the property is still under construction. Most state regulatory commissions do not permit CWIP, but, however, require utilities to accrue allowance for funds used during construction (AFUDC) and add it to the rate base only when the facility goes into operation.

Continental Power Exchange - Continental Power Exchange or CPEX is a real-time information network system for trading next-hour energy. To date, it is the only functional real-time electronic energy trading system that mirrors the Federal Energy Regulatory Commission's comparability requirements.

Contract Unit - Generally refers to the total, overall megawatt hours of a particular electricity futures contract. Per the New York Mercantile Exchange's (NYMEX) electricity futures contracts, a contract unit is 736 megawatt hours delivered over a monthly period. Per NYMEX's Permian Basin Natural Gas Futures Contract, a contract unit is 10,000 MMBtu delivered rateably over the delivery month.

Conventional Financing - Standard and accepted methods for capital project financing. Conventional financing includes both balance sheet and project financing, but does not apply to non-recourse forms of project financing.

Cost of Capital - The threshhold interest rate and return on equity that a utility or other enterprise must offer prospective investors in order to attract incremental capital.
Covenants - Provisions of debt instruments that limit actions of the borrower that might increase the risk of non-performance. An example of a debt instrument is a bond.
Credit Risk - Credit exposure itself brings about credit risk because there is an inherent risk that the counterparty to a physical (or paper) transaction may default on a deal when the contract is unfavorable to the counterparty for any number of reasons. There are three basic counterparties that present credit risk to energy marketers. They are: suppliers, customers, and financial institutions.
Cross Rate - The rate of exchange between two currencies calculated by referring to the rates between each and a third currency.
Cross Subsidization - This is the practice of transferring assets or services from the regulated portion of an electric utility to its unregulated affiliates to produce what amounts to an unfair competitive advantage. Cross subsidization also can refer to one rate class—such as industrial customers—subsidizing the rates of another class—such as residential customers.
Current Dollars - In relationship to past or future costs and prices, figures that are not adjusted for inflation. Data for each year are stated at the dollar value that are then, or are expected to be, current.
Curve Risk - The forward price for a product is the price one agrees today to buy or sell a specific amount of a commodity in the future. Forward prices are affected by the supply and demand expected during that period of time. The establishment of market prices for delivery in the future creates a forward price curve. The forward curve may be described in one of two ways: contango, when forward prices are higher than the spot price, or backwardated, when forward prices are lower than the spot price.

D

Death Spiral - A so-called death spiral occurs when a company attempts to respond to a price-induced loss in sales by charging higher prices in an attempt to maintain revenues. This action often prompts further load loss.
Debt/Equity Ratio - The relative proportion of debt and equity comprising the overall capital structure of a company.

Debt Service - The necessary outlay of funds to meet all interest and principal payments during a given year.

Default - Failure to make a financial obligation.

Delivery Period - Generally refers to the period in which delivery of a commodity contract may occur. Per the New York Mercantile Exchange's electricity futures contracts, non-holidays, 16 on-peak hours: the hour ending 0700 prevailing time to the hour ending 2200. The delivery period may be amended by time of delivery by mutual consent of the buyer and seller.

Delivery Rate - Generally refers to the method of delivery of a commodity contract. Per the New York Mercantile Exchange's (NYMEX) electricity futures contracts, the delivery rate refers to firm delivery, which is 2 megawatt-hours throughout every hour of the delivery period. This, according to NYMEX, may be amended upon mutual agreement of the buyer and seller.

Deflation - A decrease in the general price level, which results from a decrease in total spending relative to the supply of available goods on the market.

Demand Charge - Refers to a customer charge for utility service that reflects the extent to which a particular customer chooses to purchase electricity or purchase a right to draw a certain maximum volume of gas at any time during the year. Customers who purchase gas or electricity on an interruptible basis do not, therefore, pay a demand charge.

Demand-Side Management (DSM) - DSM programs refer to utility programs that affect the timing or amount of customer electricity usage. These may include energy efficiency programs targeted at reducing the energy required to serve customer needs and programs that shift electricity demand to reduce peak loads or to make more economic use of utility resources.

Department of Energy (DOE) - The DOE is a federal agency that manages the programs of research, development, and commercialization for various energy technologies and associated environmental, regulatory, and defense programs. DOE promulgates energy policies and acts as a principal adviser to the President of the United States on energy matters.

Depletion - The value loss incurred in connection with the exhaustion of the natural gas reserves in the course of service.

Depreciation - A reduction in the value of capital goods due to wear or obsolescence.

Derivatives - Consist of risk-management financial tools whose values depend upon those of their "underlying" assets. Derivatives examples include futures contracts, options, and swaps. Most usually, derivatives represent ways to reduce or control an investors' risk; however, in some cases they do not. Derivatives are often referred to as so-called hybrid investments because they don't actually represent an investment in a "real" product. For example, a crude oil futures contract is a bet on which way prices will move; what occurs to the oil itself is of small interest to the investor.

Direct Billing - Refers to lump-sum charges that are unrelated to the current level of service taken by the customer being billed.

Discount - Interest that is withheld when a note, draft, or bill is purchased.

Discount Rate - The rate of interest charged by the Federal Reserve on loans it makes to member banks. This rate has an influence over the rate banks charge customers.

Dispatchability - Describes the ability of a generating unit to increase or decrease generation or to be brought on-line or shut down at the request of a utility's system operator.

Distribution - Delivery of electricity from the transmission system to the end-user and regulated by state governments.

Dividend - The payment per share that a corporation distributes to its stockholders as their return on the money they have invested in its stock.

Downside Risk - The probability that the price of an investment will fall.

Dry Natural Gas Production - Marketed production less extraction loss.

Dual-Fuel Capacity - This is the ability of an energy customer, usually large industrial and electric utility customers, to utilize two or more kinds of fuel.

Dual Trading - The practice of allowing the same individuals to trade for their own account while at the same time executing orders for customers as a floor broker.

E

Economic Bypass - This is the bypass of existing capacity which reduces the total resource cost of delivery taking into account the stranded costs of idled capacity.

Electric Utility - An electric utility is a company that controls the distribution of electricity in a given state, area, or region. Electric utilities often own and operate electricity generation and transmission facilities. Utilities most generally in history have been afforded monopoly status—that is they have always had well-defined service territories with no competitition allowed—however, this is quickly changing. The Federal Energy Regulatory Commission on April 24, 1996 adopted two electric utility industry restructuring rules, FERC Order 888 and FERC Order 889.

Electric Utility Affiliate - An affiliate of an electric utility is a subsidiary (or "arm") of an electric utility. Many electric utilities develop, own, and operate independent power facilities, for instance, which are referred to as affiliates.

Electric Utility Consumption - Gas used as fuel in electric utility plants.

Electronic Bulletin Board (EBB) - An electronic posting system, which, among other things, gives vital information concerning energy sales and purchases, including price, terms, and conditions under which sales and purchases will occur.

Electricity Futures Contract - Offered by the New York Mercantile Exchange (NYMEX) beginning March 1996, the electricity futures contract gives utilities a market to price electricity up to 18 months into the future. The delivery points for the first contract will be on the West Coast at the California-Oregon Border (COB). A second contract has been proposed with Palo Verde, Arizona as the delivery point.

Eminent Domain - The right of a government or government authority such as a public utility to appropriate private property for public use. When invoking eminent domain, the government or government entity usually compensates the private owner of the property.

End-User - The ultimate consumer of energy. An end-user is the opposite of a customer purchasing energy for resale.

Energy - Output of generating plants in kilowatt-hours. Energy price does not include the cost of delivery. Under conditions which currently exist due to deregulation, "energy" refers only to energy as a commodity and not the ability to provide it on demand.

Energy Policy Act of 1992 (EPAct) - EPAct is the first comprehensive federal energy law promulgated in more than a decade. The act was designed to promote economic development and the wise use of energy and to encourage greater wholesale competition in the electric utility industry. The Federal Energy Regulatory Commission is the governmental agency given the greatest role in implementing EPAct, especially in the areas of approving licenses for non-utility generators and in ordering utilities to open power lines to wholesale suppliers of electricity. State regulatory agencies also play a role in implementing certain portions of the act, including retail transactions. Although the act is far-reaching in its modifaications of the existing energy market, it includes several major sections which pertain to electric utilities in general. These include: the generation of electricity; the transmission of electricity; global climate change; electric vehicle technology and research; wise use of energy; and electric and magnetic fields research.

Environmental Protection Agency (EPA) - The federal Environmental Protection Agency (EPA) was established under the Clean Air Act Amendments of 1970 to undertake all administrative and regulatory functions related to the prevention, control, and abatement of air pollution.

Equity - The value of property beyond the amount that is owed on it.

Ex Parte - As relates to regulatory procedure, ex parte refers to contacts between regulators and interested parties outside of formal proceedings. Such contacts are often circumscribed or are prohibited.

Exchange for Physicals (EFPs) - When two market participants agree to transfer, for example, title to natural gas, by mutual consent in exchange for closing out their position on the exchange.

Exchange Traded - Standardized commodities or derivatives that are traded on formally organized exchanges, such as natural gas futures and futures options on the New York Mercantile Exchange or NYMEX.

Exempt Wholesale Generators (EWGs) - Came about with the National Energy Policy Act of 1992 (EPAct), which revised the 1935 Public Utility Holding Company Act. EWGs, which are independent power facilities that generate electricity for sale in wholesale power markets, can be owned by utilities, utility holding companies, or by developers not affilated with an electric utility. EPAct freed up the existing power generation market, allowing EWGs to compete with experienced utilities already in the marketplace. The Federal Energy Regulatory Commission is responsible for determining EWG status.

Exports - Natural gas deliveries out of the Continental United States and Alaska to foreign countries.

Extraction Loss - The reduction in volume of natural gas resulting from the removal of natural gas liquid constituents at natural gas processing plants.

F

Federal Energy Regulatory Commission (FERC) - The Federal Energy Regulatory Commission is the chief energy regulatory body of the U.S. government. FERC was given new powers when the Congress passed the Energy Policy Act of 1992. Under the act, FERC is responsible for determining Exempt Wholesale Generator (or EWG) status, and has the authority to order utilities to provide access to their power transmission systems to other electric generators. FERC also certifies Qualifying Facilities or QFs as defined by the Public Utility Regulatory Policies Act or PURPA; establishes and enforces rates for power sales and transmission services; issues licenses for hydro-electric projects; and regulates aspects of mergers and acquisitions of natural gas and electric utility companies. FERC, in addition, establishes and enforces rates related to the sale and transportation of both oil and natural gas.

Federal Environmental Impact Statement (FEIS) - A statement issued by the Federal Energy Regulatory Commission in 1996 that addresses concerns in the electric power industry over the impacts that deregulation in the industry would have on the environment. In the FEIS, the FERC states that the effect of deregulation and the ensuing changes that go along with it would have only minor (if any) impact on the environment.

Firm Power - In the electric utility industry, firm power refers to power that is, through a written contract, set aside for a specific use or purpose. Under a firm contract, the power must be available at all times during that committed time frame. Curtailment is not expected nor should it be under a firm power contract.

Firm Service - Service offered with a guarantee of delivery. Customers generally pay more for firm gas than for interruptible gas, which the utility may curtail at its discrection without liability.

Fiscal Year - The 12-month period that a corporation or governmental body uses for bookkeeping purposes.

Fixed Asset - Items which cannot be quickly converted into cash without severe repercussions to business. This may include a power plan, land, or generating equipment.

Fixed Costs - Also known as embedded costs, fixed costs entail that portion of the total cost of any business activity that cannot be reduced by lowering the level of service.

Flared - The volume of gas burned in flares on the base site or at gas processing plants.

Force Majeure - A condition permitting a company to depart from the strict terms of a contract because of an uncontrollable event or effect.

Forward Contract - A contract for future delivery at a price determined in advance.

Franchise Monopoly - Describes the system in which U.S. electric utilities have operated for nearly a century. This monopoly status, however, is fading fast with the adoption of FERC Order 888 and FERC Order 889. Under a franchise monopoly, utilities have the right to be the sole or principal supplier of electric power at the retail level in a specified state, region, or area, which is known as the franchise service territory. In return for its sole supplier privilege, utilities have the obligation to serve anyone who requests services and agrees to be accountable to state and federal regulatory bodies that regulate utility performance, accounting procedures, pricing structures, and plant permitting and siting.

Functional Unbundling - A term coined by the Federal Energy Regulatory Commission in its notice of proposed rulemaking on transmission open access and made final in FERC Order 888—the order that formally deregulates the electric power industry. Functional unbundling requires public utilities to functionally unbundle their wholesale transmission services. This means that public utilities must quote separate prices for wholesale generation and transmission, take transmission service for wholesale power sales and purchases under its own tariffs that are the same price as competitors and are under the same terms and conditions of service as competitors. They also must get information about their transmission systems for their own wholesale power transactions in the same way as its competitors do (namely through a same-time information network). *See unbundling.*

Futures - These are contracts whereby an agreement is made to deliver a quantity of goods, generally commodities, at a specified price at a certain time in the future. Futures exchanges allow parties to transfer price fluctuations risks from people who don't want the risk to speculators who are willing to take on those risks. Those taking on the risks have the potential of earning large profits.

Futures Exchanges - There are 11 futures exchanges in the United States. They are the Chicago Board of Trade, the Chicago Mercantile Exchange, the Mid-America Commodity Exchange, the Philadelphia Board of Trade, the Minneapolis Grain Exchange, the Commodity Exchange (New York), the New York Cotton Exchange and its Financial Instrument Exchange, the Coffee, Sugar and Cocoa

Exchange, the New York Mercantile Exchange, the New York Futures Exchange, and the Kansas City Board of Trade.

G

General Rate Case (GRC) - A regulatory proceeding that may be either at scheduled or unscheduled intervals in which the non-fuel components of a utility's rates are determined by reference to its actual or estimated costs for a given test year.
Gigawatt (GW) - A unit of electric power equal to one billion watts, or one thousand megawatts.
Grantor - The grantor, which also is know as the writer, is the seller of an option in a futures or physicals transaction.
Greenfield Plant - Refers to a new electric power generating facility built from the ground up.
Grid - The power grid refers to the network of high-voltage transmission lines along which power moves. In the United States, there are three distinct electric power grids. They are: the Eastern Interconnection, the Texas Interconnection, and the Western Systems Coordinating Council. In addition, certain regions of the United States import electric power from the Canadian Grid. The North American Electricity Reliability Council (NERC) estimates that on the North American Continent, the electric power transmission network consists of approximately 147,000 miles of lines operating at a minimum of 230 kilovolts. Electric utilities own more than 80 percent of the North American electric power transmission network.
Gross Withdrawals - Full well stream volume, including all natural gas plant liquid and non-hydrocarbon gases, but excluding lease condensate. Gross withdrawals also include amounts delivered as royalty payments or consumed in field operations.

H

Hedging - A popular method of selling for future delivery whereby a dealer is protected from declining prices between the time the dealer buys a product and the time it is resold. Hedging strategies for commodities such as natural gas, and now for electricity, inlcude the purchase and sale of futures contracts and other such derivatives.
Hinshaw Pipeline - A pipeline or local distribution company that has received exemption by Section 1(c) of the Natural Gas Act (NGA)

from regulations pursuant to the NGA. These companies transport interstate natural gas not subject to regulations under NGA.

Holding Company - A company whose principal assets are the securities it owns in companies that actually provide goods or services. The Public Utility Holding Company Act sets a strict definition of a holding company for electric utilities.

Hydro-Electric Pumped Storage - This is a process for storing electric energy in the form of water held in an upper reservoir. Water is pumped from a lower reservoir into an upper reservoir during off-peak periods. It is then released through hydro-electric turbines into the lower reservoir for generation during peak-demand periods for management of electric system loads or to help provide needed power during system emergencies.

I

Imports - Natural gas received in the Continental United States (including Alaska) from a foreign country.

Independent Power Marketer - Also referred to as IPMs, these are power marketers that are not classified as Affiliated Power Marketers. IPMs are generally not affilated with utilities, but rather independent entities.

Independent Power Producers - Independent Power Producers (IPPs) have been a major driving force in the evolution of competitive power markets for more than a decade. IPPs generate electric or thermal energy for sale to a utility or industrial user, usually under long-term power contracts. IPPs supply about 50,000 megawatts of capacity or seven percent of the nation's power.

Independent Producer - In the natural gas industry, any person who is engaged in the production or gathering of natural gas and who sells natural gas in interstate commerce for resale but who is not engaged in the transportation of natural gas (other than gathering) by pipeline in interstate commerce.

Industrial Consumption - Natural gas used by manufacturing and mining establishments for heat, power, and chemical feedstock.

Industrial Customers - A category of energy users embracing mostly mining and manufacturing. In addition, users of natural gas for electricity generation and cogeneration by entities other than electric utilities are often classified as industrial customers.

Insider Trading - Occurs when corporate officers purchase or sell stock in their own company. Insider trading is monitored by the Securities Exchange Commission (SEC). Insider trading, in general, so long as certain rules are followed, is perfectly legal. The practice is considered by the SEC to be illegal when corporate officers use their awareness of potential problems or events in a merger or acquisition deal to affect the price of the company's stock. If the trading is manipulated to profit from the information before it is released to the public, then the trading is considered by the SEC to be illegal.

Integrated Resource Planning (IRP) - IRP is a process by which an electric utility plans for its future resource needs. Key characteristics of IRP include a long-term forecast of power needs; a comprehensive evaluation of all resource options, both supply-and-demand side; and public review of the process.

Interstate Companies - Natural gas pipeline companies subject to Federal Energy Regulatory Commission jurisdiction.

Interruptible Gas Service, Interruptible Power Service - A gas or power supply that may, at the discretion of the energy supplier, be curtailed or otherwise interrupted during emergencies or supply shortages so that service may be obtained for domestic customers. Generally, commercial and industrial customers are subject to interruptible energy supply.

Intransit Deliveries - Redeliveries to a foreign country of foreign gas received for transportation across U.S. territory and deliveries of U.S. gas to a foreign country for transportation across its territory and redelivery to the United States.

Intransit Receipts - Refers to receipts of foreign gas for transportation across U.S. territory and redelivery to a foreign country and redeliveries to the United States of U.S. gas transported across foreign territory.

Intrastate Companies - Companies not subject to Federal Energy Regulatory Commission authority.

Investor-Owned Electric Utility (IOU) - An IOU is a form of electric utility that is owned by a group of investors. Shares of IOUs are traded on public stock markets.

K

Kilowatt (kW) - One kilowatt equals 1,000 watts. A watt is a measure of the rate at which electricity is generated or consumed.

Kilowatt-Hour (KWh) - One thousand kilowatt-hours equals one megawatt-hour. A kilowatt-hour measures the quantity of electricity generated or consumed in one hour.

L

Lease and Plant Fuel - Natural gas used in well, field, and lease operations and as fuel in natural gas processing plants.
Leverage - When an investor invests a small amount of money in a deal that has much greater value than his investment. Leverage can, in some cases, actually increase an investor's risk because if the price of whatever commodity has been invested in plummets substantially, the investor actually loses a large sum of money. In some cases, it works the other way around. If the price of the invested-in commodity happens to increase sharply, the investor will realize a huge return on his initial small investment.
Lien - A company's right to take or hold the property of a debtor as security or in payment of a debt.
Life-Cycle Costing - Refers to a technique employed to evaluate the merits of a particular investment by which consideration is given to the costs and the benefits of that investment over its entire serviceable, usable life.
Lifeline Rates - A ratemaking arrangement under which a minimum amount of gas or electricity found to be essential to life is provided to residental customers at rates below average system costs. Lifeline rates are subsidized or offset by higher rates that utilities impose on residential consumption above the minimum or on other customer classes.
Liquefied Natural Gas - Natural gas that has been liquefied by reducing its temperature to minus 260 degrees Fahrenheit at atmospheric pressure.
Liquidity - The ease of finding other interested buyers or sellers for futures' contracts.
Load Factor - The ratio of average-to-peak day use calculated over the course of a specified amount of time (day, month, or year).
Load Management - Refers to techniques used by utilities to reduce wide daily and seasonal fluctuations in customer demand and thereby improve load factors.
Local Distribution Company (LDC) - An LDC is a local or region-wide public utility that retails purchased natural gas to end-users. LDCs are usually regulated by a state or provincial public utility commission.
Low-Priority Users - These users are customers that are noted in utility curtailment schedules whose service is the first to be affected by a supply shortfall. Most generally, large industrial plants and electric utilities with alternate-fuel capability are classed in the lowest priority.

M

Marginal Users - These are users of energy that by and large are large industrial and electric utility customers for which fuel-switching between gas, residual oil, or coal poses little difficulty. Marginal users represent the most price-sensitive gas consumers. Their costs for procuring alternative fuels have a large effect on the the market price of gas.

Market-Clearing Price - The price at which supply and demand are in balance with respect to a particular commodity at a particular time. A market-clearing price is high enough to prevent a shortage, but still yet low enough to ensure that all supplies then available can be sold. The market-clearing price of bulk power is determined by the interaction of supply and demand, and the cost of transmitting energy between two market areas (third-party wheeling costs). A spatial equilibrium is reached when supply and demand are satisfied simultaneously. Market clearing prices emerge as each of the market areas in the interconnected system attempts to maximize its gains.

Major Interstate Pipeline Company - A company whose combined sales for resale, and gas transported interstate, or stored for a fee, exceeded 50 million thousand cubic feet in the previous year.

Market Power - The ability to influence prices. Generally, market power is a market-based mechanism for pushing prices above variable cost.

Marketed Production - Gross withdrawals less gas used for repressuring, quantities vented and flared, and non-hydrocarbon gases removed in treating or processing operations. Includes all quantities of gas used in field and processing operations.

Megawatt (MW) - A unit of electric power equal to one million watts, or 1,000 kilowatts.

Merchant Plant - Power plants that are outside electric utility rate bases and are without power purchase agreements. Merchant plants are built to sell power in an unregulated wholesale power market without long-term power purchase agreements.

Moody's Investors Service - Issues credit ratings on corporate and municipal debt securities.

N

National Energy Board (NEB) - The NEB is Canada's counterpart to the U.S.'s Federal Energy Regulatory Commission (FERC). The NEB regulates rates and service on international, interprovincial, and frontier gas pipelines and licenses energy imports and exports.

National Futures Association (NFA) - A non-governmental trade group used by U.S. futures exchanges to enforce regulations that apply to futures trading activity.

National Independent Energy Producers (NIEP) - An association of companies that build, own, and operate independent power generating facilities, selling electricity in competitive power markets and steam to industrial companies. NIEP was formed in 1986 with the mission of creating and preserving a legislative and regulatory climate in which a competitive, environmentally-responsible, independent electricity industry can flourish.

Native Gas - Gas in place at the time that a reservoir was converted to use as an underground storage reservoir as in contrast to injected gas volumes.

Native Load - Refers to the demand of the customers of a utility located within its service territory. Typically the utility has exclusive authority to serve its so-called native load. Deregulation of the electric utility industry will change this.

Natural Gas Act (NGA) - The Natural Gas Act, which was enacted into law in 1938, regulates the sale and transportation of natural gas in interstate commerce and for other purposes.

Natural Gas - A mixture of hydrocarbon compounds and small quantities of various non-hydrocarbons existing in the gaseous phase or solution with oil in natural underground reservoirs at reservoir conditions.

Natural Gas Policy Act of 1978 (NGPA) - The Natural Gas Policy Act (NGPA) was signed into law on Nov. 9, 1978 as a framework for the regulation of most facets of the natural gas industry.

New York Mercantile Exchange (NYMEX) - The New York Mercantile Exchange (NYMEX) is the third largest U.S. futures exchange. It was founded in 1872. In early 1996, NYMEX launched an electricity futures contract designed to service the emerging competitive energy market.

Non-Hydrocarbon Gases - Typical non-hydrocarbon gases that may be present in reservoir natural gas are carbon dioxide, helium, hydrogen sulfide, and nitrogen.

North American Electric Reliability Council (NERC) - Formed in 1968 in the aftermath of the Nov. 9, 1965, blackout that affected the Northeastern United States and Ontario, Canada, the North American Reliability Council (NERC) has as its mission that of promoting the reliability of the electricity supply for North America. NERC is a not-for-profit corporation whose owners are nine Regional Councils. They follow: East Central Area Reliability Coordination Agreement (ECAR); Electric Reliability of Texas (ERCOT); Mid-Atlantic Area Council (MAAC); Mid-America Interconnected Network (MAIN); Mid-Continent Area Power Pool (MAPP); Northeast Power Coordinating Council (NPCC); Southeastern Electric Reliability Council (SERC); Southwest Power Pool (SPP); Western Systems Coordinating Council (WSCC). Alaska Systems Coordinating Council is an affiliate member.

Non-Utility Generator (NUG) - A facility that produces electric power and then sells it to an electric utility most usually under a long-term contract. NUGs may also sell thermal energy and electricity to nearby industrial customers.

O

Onsystem Sales - In natural gas, onsystem sales refers to customers where the delivery point is a point on, or directly interconnected with, a transportation, storage, and/or distribution system operated by the reporting company.

Open Access - Access to the commodity market through unbundled transmission capacity on substantially equal terms for all. The Federal Energy Regulatory Commission (FERC) used the term open access during the late 1980s and early 1990s as shorthand for the various competitive reforms it was promoting in the natural gas industry. FERC also is currently using the term "open access" as the electric utility industry undergoes deregulation.

Open Access Same Time Information System (OASIS) - Refers to FERC Order 889, which is known as the OASIS rule. The rule ensures that transmission owners and their affiliates do not have an unfair competitive advantage in using transmission to sell power. Under the rule, public utilities must obtain information about their transmission system for their own wholesale power transactions in the same way that their competitors do—through an OASIS on the Internet. In addition, utilities must separate their wholesale power marketing and transmission operations from one another.

Options - A financial instrument (derivative) that gives the buyer the right to purchase or sell a security at a predetermined price. There are two types of options: call options and *put options.* Call options refer to the right a purchaser of a commodity has to buy the commodity at a specified price and date. A put option refers to the right of the seller of a commodity to sell the commodity at a specified price and date. Options having positive intrinsic value are said to be in-the-money options. For calls, in-the-money options are those options with a strike price below the current market price of the underlying instrument. The option holder who chose to exercise his option would be able to buy at a lower price and sell at a higher price. In-the-money put options have strike prices above the market price for the underlying instrument. Exercise of this option, would allow the holder to sell at a higher price, through the option, and to buy at a lower price. Generally, options that are in-the-money as of the expiration date will be exercised due to the fact that it is profitable. At-the-money or out-of-the-money options refer to those options that have a strike price that is the same as the current market price. At-the-money options have no intrinsic value.

Over-The-Counter - Refers to a financial asset or commodity that is bought and sold away from an organized exchange. It is thus an asset over which parties negotiate and conduct transactions directly between themselves. In natural gas markets, for example, basis and time swaps are all bought and sold over the counter.

P

Peak Day - On an annual basis peak day is the day of highest customer demand.

Performance Based Rates - Rates based on standards of service and reliability with incentives to improve service and lower costs.

Performance-Based Ratemaking (PBR) - PBR is intended to reward regulated companies on the basis of independent standards of price or productivity changes, for example, comparison of their prices with those of other utilities, or with the general price level.

Performance Criteria - Performance criteria are built-in components of a power purchase agreement to assure the purchasing utility that an independent energy producer can meet the utility's requirements at the time they are needed. Generally, performance criteria set standards for the reliability of a facility in terms of actual kilowatt-hours of output or hours of availability.

Physical Risk - In commodity trading, physical risk generally refers to the risk that the physical, tangible commodity will not be a suitable grade.
POOLCO - Describes a centrally-dispatched pool—much like that used in the United Kingdom—where transparent real-time spot pricing emerges. The U.K. requires that all generation be bought and sold within the pool. Another definition of POOLCO, offered by Southern California Edison Co., which supports the concept, is a privately-owned business, independent of all utilities and generators that assures access to the transmission grid by all sellers. The function of the POOLCO would be to monitor a region's electricity needs and then link power producers with buyers. The independent entity would also dispatch generation and transmission based upon need, provide price visibility to all buyers and sellers, and assure reliable system operation and adquate power supply. In California, some hail the POOLCO as a way to establish a competitive wholesale market where all buyers benefit from competition between generators. Those favoring POOLCO generally are those utilities that generally have high generation costs. Those in opposition include consumer groups, free-market economists, and utilities that have the lowest average generating costs. Opponents claim that the POOLCO wrongly attempts to predict and plan a single electric future for the state of California and merely spells out more of the same thing that is currently going on—central planning and so-called "micro-management."
Power Purchase Agreement - An agreement entered into between a buyer and a seller of energy. The agreement specifies the terms and conditions under which electric power will be generated and purchased. Power purchase agreements require the seller (the producer) to supply power at a specified price for the life of the agreement. While power purchase agreements vary, they do have some common elements. They are: specification of the size and operating parameters of the generating facility; milestones, in-service dates and contract terms; price mechanisms; service and performance obligations; dispatchability options; and conditions of termination or default.
Pipeline Fuel - Gas consumed in the operation of pipelines, primarily in compressors.
Power Broker - Power brokers, in contrast to power marketers, are not currently regulated by the Federal Energy Regulatory Commission. Power brokers, like power marketers, arrange bulk power transactions. However, unlike power marketers, power brokers

bring together a willing seller and buyer without taking title to the power. Thus, power brokers, unlike power marketers, assume no risk.

Power Marketer - Power marketers, like power brokers, arrange bulk power transactions. The difference, however, is that power marketers actually take title to the power—and thus the risk—associated with "owning" the power for a time. Power marketers, once taking title to the power and "owning" it for a time, resell it on the open market.

Premium - Refers to the price of an option. The value of an option has two components: intrinsic value and time value. Intrinsic value refers to the value to the holder if the option was exercised immediately. Time value refers to the value to an option holder that comes from the potential for favorable changes in the market price before the expiration of his option.

Price Elasticity of Demand - A measure of the sensititvity of demand to changes in price. This specifically means the ratio between the percentage change in sales volume and the percentage change of the price change which causes the change in sales.

Price Risk - The exposure a company has to changes in the spot or forward price of a given commodity.

Pro-Forma Tariff - Refers to the open access transmission tariffs in FERC Order 888, the major electric utility industry restructuring rule. Under the pro-forma tariff, all public utilities that own, control, or operate interstate transmission facilities are required to offer service to others. They must also use the pro-forma tariffs for their own wholesale energy sales and purchases.

Project Financing - Refers to the most common method used to finance the construction of independent power facilities. Typically, the developer pledges the value of the plant and part or all of its expected revenues as collateral to secure financing.

Public Utility - A business that may be either privately or governmentally owned. Public utilities are subject to governmental regulation because they are deemed to be a business "affected with a public interest" and to provide an "essential service" to the consuming public.

Public Utility Commission (PUC) - An administrative or quasi-judicial body at the state provincial or municipal level, whose functions include regulation of public utilities.

Public Utility Holding Company Act (PUHCA) - This act was passed by Congress in 1935 to restrain activities of large holding companies in the gas and electric utility industries. Currently, changes in the energy industry are causing legislators, regulators, and other industry stakeholders to question if this legislation has outlived is time. PUHCA vests in the federal Securities and Exchange Commission authority over those who hold interests in utilities and restricts the operations of the registered public utility holding company systems. PUHCA imposes limitations on the structure and activities of a company that directly or indirectly owns or controls 10 percent or more of a public utility. Any person who exercises a controlling influence over the management or policies of a holding company or public utility is subject to SEC regulation under terms of the act. There are two types of holding companies: registered and exempt. Registered refers to those holding companies that are fully regulated under the constraints of PUHCA. Exempt refers to holding companies that are subject to state regulation, but not special SEC requirements. The only exception to this is that exempt holding companies cannot own utility company interests outside their geographic area, nor can they acquire more than five percent of a utility's security without SEC consent.

Public Utility Regulatory Policies Act (PURPA) - Enacted by Congress in 1978, PURPA's goal initially was to help reduce the United States' dependence on foreign oil. The act encourages the use of renewable energy resources to generate electricity and increase the efficiency of fossil fuel electric generation. Section 210 encourages private investors and individuals to build small electric power production facilities using such renewable energy resources as wind, sun, solid waste, and run-of-the-river hydroelectric, etc. The act also encourages industrial plants to cogenerate electricity by using excess steam from manufacturing processes.

Put Option - A put option is the right to sell a commodity, either futures or physicals, at a specified price by a certain date. *Compare call option.*

Q

Qualifying Facility (QF) - Created with the passage of the Public Utility Regulatory Policies Act (PURPA) by Congress in 1978. To be considered a qualified facility under PURPA, a facility must meet one of three criteria: it must be a cogenerator; it must be an eligible solar, wind, waste, or geothermal facility; or it must produce 80MW or less burning bio-mass or use hydroelectric sources. There are two types of QFs. They are known as Qualifying Small Power Producers and Qualifying Cogeneration Facilities. Small Power Producers use renewable energy sources, such as wind, solar, gothermal, hydro, biomass, or waste resources. PURPA specified a size limit of 80MW on Small Power QFs, but this limit was waived by Congress in 1990 for all facilities except hydro-power plants that requested QF status by Dec. 31, 1994. A Qualifying Cogeneration Facility, the second type of QF, must produce both useful steam and electricity and must meet PURPA-specified efficiency standards if it is fueled by oil or natural gas. PURPA also specifies that a QF cannot be more than 50 percent owned by or controlled by an electric utility or an electric utility holding company. Power producers meeting the QF requirements of PURPA are exempt from many federal and state regulations governing electric utilities. For example, many QFs are generally free from rate regulation and oversight of financing instruments; however, they are subject to the same environmental regulations as utility plants.

R

Rate Base - A utility's rate base is the value of its property as recognized by a regulatory authority upon which the company is permitted to earn its rate of return.

Rate Design - A utility's rate design consists of the rules and methodology used to convert a utility's revenue requirement into specific rates for specific categories of customers and service.

Ratepayer - Refers to a retail consumer of the electricity distributed by an electric utility. This includes residential, commercial, industrial, and wholesale users of electricity.

Reengineering - Reengineering is a term popularized by management consultant Michael Hammer. It generally involves the redesign of specific business processes to achieve major improvements in critical, contemporary measures of performance, which can include cost, quality, service and speed.

Regional Transmission Group - A voluntary organization of transmission owners, transmission users, and other entities interested in coordinating transmission planning and expansion, operation and use on a regional and inter-regional basis.

Regulatory Compact - In the electric utility industry, a metaphorical description of the industry that establishes an electric utility's duty to serve consumers, an electric utility's right to cost recovery and the utility's obligation to take service. This so-called "regulatory compact" took for granted that the vertically integrated monopoly of electric utilities would forever be efficient servers. The compact got its beginnings during the period of a perceived need to suppress inefficient competition and replace it with efficient monopoly. The compact metaphor contains no provisions for what should happen if efficient competition threatens a relatively ineffecient monopoly utility. The regulatory compact that the electric utility industry has enjoyed for nearly a century is quickly coming to a close with deregulation of the industry.

Regulatory-Out Clause - A clause contained in a power purchase agreement which permits a utility to shift its regulatory risk to the owner of an independent power project.

Renewable Energy - Energy characterized as renewable is any energy source that is constantly replenished through natural processes. Sunlight, moving water, geothermal springs, biomass, and wind are all examples of renewable energy resources used to generate electricity.

Repowered Plant - An existing power plant that has been rebuilt to extend its useful life.

Repressuring - The injection of gas into oil or gas formations to effect greater ultimate recovery.

Residential Consumption - Gas used in private dwellings, including apartments, for heating, cooking, water heating, and other household uses.

Residential Customer - The category of energy users that consists of individually-billed households.

Retail Wheeling - The transmission of power to an individual customer from a generator of electricity other than the host utility. The National Energy Policy Act, enacted in 1992, prohibits the Federal Energy Regulatory Commission from mandating retail wheeling. States and their regulatory bodies, however, are free to enact their own retail wheeling initiatives. Many have. Several federal retail wheeling bills have been introduced and are under investigation at the federal level. Several congressmen feel that although the states should still retain their regulatory authority over state retail wheeling activity, federal legislation needs to be enacted to avoid a patchwork quilt of retail wheeling regulations nationwide.

Retrofit - A retrofit is a process by which an existing piece of equipment or a whole facility is changed or upgraded to improve its efficiency and/or performance.

Risk - In the energy industry, there are five categories of risks: price, basis, curve, physical, and credit. See separate entries for definitions under price risk, basis risk, physical risk, and credit risk.

Risk Management - Risk management refers to tools, such as those offered to the energy industry by the New York Merchantile Exchange for natural gas and electricity, to strike a balance between risk mitigation and risk taking that is consistent with a company's overall risk tolerance in the commodities market. "Hedging" is an example of a risk management tool. Another example is derivatives.

S

Securities and Exchange Commission (SEC) - Created by the U.S. government in 1934 following the Great Depression and the stock-trading scandals that it exposed. The SEC's mission is to regulate stock traders' activities. The SEC is comprised of five members, who are appointed by the President. Members of SEC oversee a team of attorneys, accountants and investigators who ensure that the securities markets operate honestly and fairly. The SEC, when necessary, can enforce securities laws with sanctions, ranging from fines to prosecution.

Service Territory - Refers to the state, area, or region served by an electric utility. Under deregulation of the electric utility industry, the boundaries that used to be associated with service territories will be blurred or maybe even removed altogether.

Stakeholders - The major players or interested parties of a utility or regulatory agency and their actions. Examples of stakeholders include: shareholders, management, employees, suppliers, customers, and competitors.

Standard & Poor's - Issues credit ratings on corporate and municipal debt securities.

Stockholder's Equity - The value of the shares a corporate stockholder holds.

Stranded Investment - Also sometimes referred to as "stranded costs," stranded investment refers to fixed costs—usually related to investment in generation facilities—incurred by a utility which would no longer be paid by customers who have opted to buy their power from other suppliers. Under FERC Rule 888, the major electric power industry restructuring rule, stranded investments can be recovered from a utility's departing customers.

Storage Additions - The volume of gas injected or otherwise added to underground natural gas or liquefied natural gas storage during the applicable reporting period.

Storage Withdrawals - Total volume of gas withdrawn from underground storage or liquefied natural gas storage during the applicable reporting period.
Straddle - In futures, straddle is also known as a spread. A straddle trade is based on the price relationship between the purchase of one futures month against the sale of another futures month of the same commodity. In options, straddle refers to the sale or purchase of a put option and a call option that have the same strike price and expiration date.
Strangle - In options, a strangle is similar to a straddle as it relates to options, only in a strangle, there are different strike prices, which are usually at out-of-the money prices.
Strike Price - Also called an exercise price, the strike price refers to the purchase price referenced in a call option or sale price specified in a put option.
Supplemental Power - This is power (energy or capacity) that an electric utility may supply to a qualified facility for in-house load requirements or outages.
Supplemental Gaseous Fuels Supplies - Synthetic natural gas, propane-air, refinery gas, biomass gas, air injected for stabilization of heating content, and manufactured gas commingled and distributed with natural gas.
Swaps - Private agreements between two parties—a buyer and seller of a physical commodity or financial asset—to exchange cash flows from physical transactions in the future based on a prearranged formula. Swaps are over-the-counter financial derivatives.
Synthetic Natural Gas - A manufactured product, chemically similar in most respects to natural gas, that results from the conversion or reforming of petroleum hydrocarbons and may easily be substituted for or interchanged with pipeline quality natural gas.

T

Take-or-Pay Clause - Generally refers to a natural gas supply contract wherein the purchaser must pay the supplier for a certain percentage of the fuel under contract, regardless of whether the purchaser actually took possession of the fuel, used the fuel, or resold it.
Take-and-Pay Clause - Generally refers to a natural gas supply contract wherein the purchaser must take physical possession of the fuel, regardless of whether the purchaser can use it or resell it.

Termination of Trading - This term is generally applied to futures contracts and refers to the time in which, within a delivery period, trading ceases. Per the New York Mercantile Exchange's (NYMEX) electricity futures contracts, termination of trading occurs on the third business day prior to the first day of the delivery month. Per NYMEX's Permian Basin Natural Gas Futures Contract, trading terminates three business days prior to the first day of the delivery month.
Therm - One-hundred thousand British thermal units.
Topping Cycle - A Topping Cycle refers to an electricity-producing cogeneration facility that captures waste heat in the process of producing the electricity. The waste heat is then used for useful thermal application.
Transition Costs - A reduction in the net present value of company assets attendant on exposure to competitive markets. These costs include any costs passed on to ratepayers during the transition period to competition in the electric utility industry. These costs will only include the amount of reduction in value that is actually recovered. In the natural gas industry, the Federal Energy Regulatory Commission in Orders 500 and 636 included provisions permitting interstate pipelines to recover some of their claimed transition costs from historical customers.
Transmission - High-voltage system to move large blocks of power from power plants to load centers. Under Federal Energy Regulatory Commission Order 888, FERC will not set rates for transmission (and ancillary services) as originally proposed in the restructuring of the electric utility industry. Public utilities, the order states, will make rate filings to recover their costs.

U

Unbundling - The separation of various industry functions for pupos-es of offering and billing for services. In the natural gas transmission and distribution sectors, unbundling refers to the separation of sales and transport activities. In the electric utility industry, unbundling refers to the separation of generation and transmission activities. *See functional unbundling.*
Underground Gas Storage Reservoir Capacity - Interstate company reservoir capacities are those certified by the Federal Energy Regulatory Commission. Independent producer and intrastate company reservoir capacities are reported as developed capacity.

Underlying Commodity - Also referred to as underlying instrument, underlying commodity is the product that is specified in an option.
Uneconomic Bypass - Bypass that yields a net benefit to the bypasser, but which benefit is less than the total costs it imposes on other parties, particularly the costs of redundant physical plant.
Utilization Factor - The ratio of the peak demand of an electrical generation system to the rated capacity of the generating system.

V

Value of Service Pricing - Also called "Ramsey pricing," value of service pricing is monopoly price discrimination in a form approved by public-utility regulators. A method of apportioning costs among cusotmers so that users who place a greater value on the service are charged higher rates than the more price-sensitive customers.
Variable Cost - Refers to the portion of a utility's cost of service which varies with the volume of sales.
Vented Gas - Gas released into the air on the base site or at processing plants.
Vertically-Integrated Utility - Utilities that sell power on a bundled basis and whose activities run the full range of different functional activities-generation, transmission, and distribution. With deregulation in the electric utility industry well under way, vertically-integrated electric utilities are on their way out.

W

Wellhead Price - Represents the wellhead sales price, including charges for natural gas plant liquids subsequently removed from the gas, gathering, and compression charges, and state production, severance, and/or similar charges.
Wheeling - The Energy Policy Act of 1992 (EPAct) requires electric utilities to allow outside wholesale electricity generators to have access to their power lines. The Federal Energy Regualtory Commission can order utilities to do so if it decides that certain conditions have been adequately satisfied. This transmission of electricity for a third party is called "wheeling." *Also see retail wheeling and wholsale wheeling.*
Wholesale Wheeling - The use of transmission facilities of one system to transmit power (energy) by agreement of and for another system

with a corresponding wheeling charge. Wholesale wheeling involves only sales for resale and occurs when the buyer of the power resells the wheeled power to retail customers.

Working (Top Storage) Gas - The volume of gas in an underground storage reservoir above the designed level of the base. It may or may not be completely withdrawn during any particular withdrawal season. Conditions permitting, the total working capacity could be used more than once during any season.

ACRONYMS

A

AAPG - American Association of Petroleum Geologists
ACA/CIPCA - American Cogeneration Association/Cogeneration & Independent Power Coaltion of America
AFC - Allowance of Funds for Construction
AGA - American Gas Association
AFBD - Association of Futures Brokers and Dealers
AFUDC - Allowance for Funds Used During Construction
AIMS - Automated Interchange Matching System, Inc.
AM/FM - Automated Mapping/Facilities Management
ANEC - American Nuclear Energy Council
ANGTS - Alaska Natural Gas Transportation System
ANS - American Nuclear Society
AOSC - Association of Oil Well Servicing Contractors
AP - Associated Person (as relates to a Futures Commission Merchant employee)
API - Application Program Interface also American Petroleum Institute
APM - Affiliated Power Marketer
APPA - American Public Power Association
ASCC - Alaska Systems Coordinating Council
ATC - Available Transmission Capacity

B

BACT - Best Available Control Technology
bcf - Billion Cubic Feet
BLS - Bureau of Labor Statistics (Department of Commerce)
BPTP - Bulk Power Transmission Protocol
BRPU - Biennial Resource Planning Update
Btu - British Thermal Unit

C

CAA - Clean Air Act
CAAA - Clean Air Act Amendments of 1990
CC Plant - Combined-Cycle Plant
CEC - California Energy Commission
CEA - Commodity Exchange Act of 1936
CEMS - Continuous Emission Monitoring System
CEQ - Council on Environmental Quality
CF - Cubic Foot
CFB - Circulating Fluidized Bed
CFDs - Contracts For Differences
CFTC - Commodities Futures Trading Commission
CIF - Cost, Insurance, and Freight Price
CLECA - California Large Energy Consumers Association
CNG - Compressed Natural Gas
CO - Carbon Monoxide
CO_2 - Carbon Dioxide
CPEX - Continental Power Exchange Inc.
CPO - Commodity Pool Operator
CPUC - California Public Utilities Commission
CRT - Capacity Reservation Tariff
CT - Combustion Turbine
CTA - Commodity Trading Advisor
CWIP - Construction Work in Progress

D

DOE - U.S. Department of Energy
DSM - Demand-Side Management

E

EBB - Electronic Bulletin Board
ECAR - East Central Area Reliability Coordination Agreement
EEI - Edison Electric Institute
EFP - Exchange of futures for physicals
EGA - Electric Generation Association
EHV - Extra-High Voltage
EIN - Electronic Information Network
EISB - Electric Industry Standards Board
EIS - Environmental Impact Statement
EOR - Enhanced Oil Recovery
EPA - Environmental Protection Agency
EPC - Engineering, Procurement, and Construction
EPRI - Electric Power Research Institute
ERAM - Electric Revenue Adjustment Mechanism
ERCOT - Electric Reliability Council of Texas
ESCOs - Energy Service Companies
EWG - Exempt Wholesale Generator

F

FCM - Futures Commission Merchant
FEIS - Federal Environmental Impact Statement
FERC - Federal Energy Regulatory Commission
FGD - Flue Gas Desulfurization
FOB - Free on Board Price
FPA - Federal Power Act
FPC - Federal Power Commission
FSA - Financial Services Act of 1986 (United Kingdom)
FV - Fixed Variable Rate Structure

G

GISB - Gas Industry Standards Board
GRC - General Rate Case
GW - Gigawatt

H

HRSG - Heat Recovery Steam Generator

I

IADC - International Association of Drilling Contractors
IB - Introducing Brokers
IEA - International Energy Agency
IEEE - Institute of Electrical and Electronics Engineers
INPO - Institute of Nuclear Power Operations
IPAA - Independent Petroleum Association of America
IPM - Independent Power Marketer
ISO - Independent System Operator
IOU - Investor-Owned Utility
IROR - Incentive Rate of Return
IRP - Integrated Resource Planning
IPP - Independent Power Producer

L

LAER - Lowest Achieveable Emission Rate
L/C - Letter of Credit
LDC - Local Distribution Company
LNG - Liquified Natural Gas
LOC - Line of Credit
LOI - Letter of Intent
L.P. - Limited Partnership
LDs - Liquidated Damages
LNG - Liquefied Natural Gas
LPG - Liquified Petroleum Gas
LRAC - Long-Run Avoided Cost

M

MAAC - Mid-Atlantic Area Council
MAIN - Mid-America Interpool Network
MAPP - Mid-Continent Area Power Pool
Mcf - Thousand cubic feet (gas)
MFV Rates - Modified Fixed-Variable Rates

MMBtu - Million Btus
MMCF - Million cubic feet (gas)
MMS - Minerals Management Service (Department of Interior)
MOU - Memorandum of Understanding
MW - Megawatt
MWh - Megawatt Hour

N

NARUC - National Association of Regulatory Commissioners
NEMA - National Electrical Manufacturers Association
NEPOOL - New England Power Pool
NERC - North American Electric Reliability Council
NFA - National Futures Association
NGA - Natural Gas Act
NGLs - Natural Gas Liquids
NGPA - Natural Gas Policy Act of 1978
NEB - National Energy Board (Canada)
NERC - North American Electric Reliability Council
NFA - National Futures Association
NGLs - Natural Gas Liquids
NGPA - Natural Gas Policy Act of 1978
NIEP - National Independent Energy Producers
NRC - Nuclear Regulatory Commission
NRDC - Natural Resouces Defense Council
NOI - Notice of Inquiry
NOPR - Notice of Proposed Rulemaking
NPCC - Northeast Power Coordinating Council
NRC - Nuclear Regulatory Commission
NRECA - National Rural Electric Cooperative Association
NRRI - National Regulatory Research Institute
NSPS - New Source Peformance Standards
NSR - New Source Review
NUG - Non Utility Generator
NSWA - National Stripper Well Association
NYMEX - New York Mercantile Exchange

O

OASIS - Open Access Same-time Information System (FERC Order 889)
OCS - Outer Continental Shelf

O&M - Operations and Maintenance
OTAG - Ozone Transport Assessment Group

P

PBR - Performance-Based Rates
PGA - Purchased Gas Adjustment
PJM - Pennsyvania/New Jersey/Maryland Interconnection
PMA - Power Marketing Association
PPA - Purchase Power Agreement
PSA - Power Sales Agreement
PSC - Public Service Commission
psi - Pounds Per Square Inch
PUC - Public Utility Commission
PUD - Public Utility District
PUHCA - Public Utility Holding Company Act
PURPA - Public Utility Regulatory Policies Act

Q

QF - Qualifying Facility

R

R&D - Research and Development
RAC - Refiners Acquisition Costs
RACT - Reasonably Available Control Technology
RCRA - Resource Conservation and Recovery Act
REA - Rural Electrification Administration
RIE - Recognized Investment Exchange
RIM - Ratepayer Impact Measure Test
RIN - Real-Time Information Network
ROC - Regulatory-Out Clause
ROE - Return on Equity
R/P Ratio - Reserves-To-Production Ratio
RTG - Regional Transmission Group

S

SCADA - Supervisory Control and Data Acquisition

SEC - Securities and Exchange Commission
SCF - Standard Cubic Feet (also CF or Cubic Foot)
SERC - Southeastern Electric Reliability Council
SFV - Straight Fixed Variable
SIB - Securities Investment Board
SIP - State Implementation Plan
SMP - Special Marketing Program
SNG - Synthetic Natural Gas
SO_2 - Sulfur Dioxide
SRO - Self-Regulatory Organization
SSA - Steam Sales Agreement
SPP - Southwest Power Pool
SSA - Steam Sales Agreement

T

TDS - Total Dissolved Solids
TOP - Take-Or-Pay
TOU - Time-of-Use Tariffs
TPY - Tons Per Year
TRC - Total Resource Cost Test
TSIN - Transmission Services Information Network

U

USCEA - U.S. Council for Energy Awareness
USEA - United States Energy Association
UTC - Utilities Telecommunications Commission

W

WACOG - Weighted Average Cost of Gas
WAPA - Western Area Power Administration
WSCC - Western Systems Coordinating Council
WSPP - Western Systems Power Pool

SELECTED TRADE GROUPS, ASSOCIATIONS AND FEDERAL AGENCIES

Automated Mapping/Facilities Management (AM/FM) International
14456 E. Evans Ave.
Aurora, CO 80014
Phone: (303) 337-0513
Fax: (303) 337-1001

American Cogeneration Association/Cogeneration & Independent Power Coalition of America (ACA/CIPCA)
1025 Thomas Jefferson St. NW
Box One
Washington, D.C. 20007
Phone: (202) 965-1134
Fax: (202) 965-1139

American Gas Association (AGA)
1515 Wilson Blvd.
Arlington, VA 22209
Phone: (703) 841-8564
Fax: (703) 841-8406

American Nuclear Energy Council (ANEC)
410 First St. SE
Washington, D.C. 20003
Phone: (202) 484-2670
Fax: (202) 484-7320

American Nuclear Society (ANS)
555 N. Kensington Ave.
La Grange Park, IL 60525
Phone: (708) 352-6611
Fax: (708) 352-0499

American Petroleum Institute (API)
1220 L St. NW
Washington D.C. 20005
Phone: (202) 682-8000
Fax: (202) 682-8115

American Public Gas Association (APGA)
11094 Lee Highway, Ste. D-102
Fairfax, VA 22080
Phone: (703) 352-3890
Fax: (703) 352-1721

American Public Power Association (APPA)
2301 M. St., NW
Washington, D.C. 20038
Phone: (202) 775-8300
Fax: (202) 467-2910

American Solar Energy Association
2400 Central Ave., B-1
Boulder, CO 80301
Phone: (303) 443-3130
Fax: (303) 443-3212

Associated Gas Distributors (AGD)
1001 Pennsylvania Ave. NW
Washington D.C. 20004
Phone: (202) 466-5829
Fax: (202) 628-5116

Association of Energy Engineers
4025 Pleasantdale Rd., Ste. 420
Atlanta, GA 30340
Phone: (404) 447-5083
Fax: (202) 446-3969

Bonneville Power Administration (BPA)
P.O. Box 3621
Portland, OR 97208
Phone: (503) 230-3000
Fax: (503) 230-4550

Cogeneration Institute
4025 Pleasantdale Rd., Ste. 420
Atlanta, GA 30340
Phone: (404) 447-5083
Fax: (404) 446-3969

Council of Petroleum Accountant Societies (COPAS)
P.O. Box 1190
Denison, TX 75021
Phone: (903) 463-5463
Fax: (903) 463-5473

Council on Environmental Quality (CEQ)
722 Jackson Place NW
Washington, D.C. 20503
Phone: (202) 395-5750
Fax: (202) 395-3744

Department of Energy (DOE)
National Energy Information Center, EI-231
Energy Information Administration
Forrestal Bldg., Room IF-048
Washington, D.C. 20585
Phone: (202) 586-8800
Fax: (202) 586-0727

Domestic Petroleum Council
c/o Bogle & Gates
1299 Pennsylvania Ave. NW, No. 875E
Washington, D.C. 20004
Phone: (202) 296-1709
Fax: (202) 293-6825

Edison Electric Institute (EEI)
701 Pennsylvania Ave., N.W.
Washington, D.C. 20004
Phone: (202) 508-5000
Fax: (202) 508-5794

Electric Generation Association (EGA)
2715 M St. NW, Ste. 150
Washington, D.C. 20007
Phone: (202) 965-1134
Fax: (202) 965-1139

Electric Power Research Institute (EPRI)
In California:
3412 Hillview Ave.
Palo Alto, CA 94304
Phone: (415) 855-2000
Fax: (415) 855-2954

In Washington
1019 19th St. NW, Ste. 1000
Washington, D.C. 20036
Phone: (202) 872-9222
Fax: (202) 296-5436

Electricity Consumers Resource Council
1333 H St. NW
West Twr 8th FL
Washington, D.C. 20005
Phone: (202) 682-1390
Fax: (202) 289-6370

Energy Council
122 W. Carpenter Freeway, Ste. 525
Irving, Texas 75039
Phone: (214) 717-8105
Fax: (214) 717-6107

Environmental Protection Agency (EPA)
401 M St. SW
Washington, D.C. 20460
Phone: (202) 260-2090
Fax: (202) 260-0279

Federal Energy Regulatory Commission (FERC)
825 N. Capital St. NE
Washington, D.C. 20426
Phone: (202) 208-0200
Fax: (202) 208-0147

Gas Appliance Manufacturers Association (GAMA)
1901 N. Moore St., Ste. 1100
Arlington, VA 22209
Phone: (703)525-9565
Fax: (703) 525-8159

Gas Industry Standards Board (GISB)
1100 Louisiana, Ste. 4925
Houston, Texas 77002
Phone: (713) 757-2288
Fax: (713) 757-2491

Gas Processors Association
6526 E. 60th St.
Tulsa, OK 74145
Phone: (918) 493-3872
Fax: (918) 493-3875

Gas Research Institute
8600 Bryn Mawr Ave.
Chicago, IL60631
Phone: (312) 399-8100
Fax: (312) 399-8170

Independent Petroleum Association of America (IPAA)
1101 16th St. NW, 2nd Floor
Washington, D.C. 20036
Phone: (202) 857-4722
Fax: (202) 857-4799

Institute of Nuclear Power Operations (INPO)
1100 Circle 75 Parkway, Ste. 1500
Atlanta, GA 30339
Phone: (404) 953-3600
Fax: (404) 953-7549

Interstate Natural Gas Association of America (INGAA)
555 13th St. NW, Ste. 300W
Washington, D.C. 20004
Phone: (202) 626-3200
Fax: (202) 626-3289

Interstate Oil and Gas Compact Commission (IOGCC)
Box 53127
Oklahoma City, OK 73152
Phone: (405) 525-3556
Fax: (405) 525-3592

Mid-Continent Oil and Gas Association
801 Pennsylvania Ave. NW, Ste. 340
Washington, D.C. 20004
Phone: (202) 638-4400
Fax: (202) 638-5967

Midwest Cogeneration Association
c/o Enerteq Sciences Inc.
P.O. Box 302
Western Springs, IL 60558
Phone: (708) 246-6808

Midwest Gas Association
7831 Glanroy Rd., Ste. 300
Minneapolis, MN 55439
Phone: (612) 832-9915
Fax: (612) 832-9308

National Association of Energy Service Companies
1440 New York Ave.
Washington, D.C. 20005
Phone: (202) 371-7980
Fax: (202) 393-5760

National Association of Regulatory Utility Commissioners (NARUC)

Box 684
1102 I.C.C. Bldg.
Washington, D.C. 20044
Phone: (202) 898-2200
Fax: (202) 898-2123

National Association of Royalty Owners

119 N. Broadway
Ada, OK 74820
Phone: (405) 436-0034
Fax: (405) 436-1535

National Drilling Contractors Association

3008 Millwood Ave.
Columbia, S.C. 29205
Phone: (803) 259-5646

National Electrical Manufacturers Association (NEMA)

2101 L St. NW
Washington, D.C. 20037
Phone: (202) 457-8400
Fax: (202) 457-8468

National Independent Energy Producers (NIEP)

601 13th Street NW/Suite 320 South
Washington D.C. 20005
Phone: (202) 783-2244
Faxt: (202) 783-6506

National Petroleum Council

1625 K Street NW, Ste. 600
Washington, D.C. 20006
Phone: (202) 393-6100
Fax: (202) 331-3539

National Regulatory Research Institute (NRRI)

1080 Carmack Rd.
Columbus, OH 43210
Phone: (614) 292-9404
Fax: (614) 292-7196

National Rural Electric Cooperative Association (NRECA)
4301 Wilson Blvd.
Arlington, VA 22203
Phone: (703) 907-5500
Fax: (703) 907-5519

Natural Gas Council
c/o Natural Gas Supply Association
805 15th St. NW, Ste. 510
Washington, D.C. 20005
Phone: (202) 326-9300
Fax: (202) 326-9330

Natural Gas Supply Association (NGSA)
805 15th St. NW, Ste. 510
Washington, D.C. 20005
Phone: (202) 326-9300
Fax: (202) 326-9330

Natural Gas Training & Education Association
11811 N. Freeway, Ste. 256
Houston, Texas 77060
Phone: (713) 447-7372
Fax: (713) 447-2642

New England Cogeneration Association
1040 Great Plain Ave.
Needham, MA 02192
Phone: (617) 444-9980
Fax: (617) 444-9981

North American Electric Reliability Council (NERC)
101 College Road East
Princeton, NJ 08540
Phone: (609) 452-8060
Fax: (609) 452-9550

Northern California Cogeneration Association
c/p Impell Corporation
350 Lennon Lane
Walnut Creek, CA 94958
Phone: (415) 943-4500

Nuclear Regulatory Commission (NRC)
Washington, D.C. 20555
Phone: (301) 492-7000

Pennsylvania Cogeneration Coalition
5 Great Valley Parkway, Ste 262
Malvern, PA 19355
Phone: (215) 251-9833

Power Marketing Association
1519 22nd Street, Ste 200
Arlington, VA 22202
Phone: (703) 892-0010
Fax: (703) 979-4677

Rural Electrification Administration (REA)
14th and Independence Ave. SW
Rm. 4037-South Agriculture Bldg.
Washington, D.C. 20250
Phone: (202) 720-9560
Fax: (202) 720-1954

San Diego Cogeneration Association
c/o Onsite Energy
701 Palomar Airport Rd. Ste 200
Carlsbad, CA 92009
Phone: (619) 931-2400
Fax: (619) 931-2405

Scientists & Engineers For Secure Energy Inc.
570 7th Ave., Ste. 1007
New York, NY 10018
Phone: (212) 840-6595
Fax: (212) 840-6597

Solar Energy Industries Association
777 N. Capitol St. NE, Ste. 805
Washington, D.C. 20002
Phone: (202) 408-0660
Fax: (202) 408-8536

Southeastern Power Administration
Samual Elbert Bldg., Public Square
Elberton, GA 30635
Phone: (706) 283-9911

Southwestern Power Administration
P.O. Box 1619
Tulsa, OK 74101
Phone: (918) 581-7420

Tennessee Valley Authority (TVA)
400 W. Summitt Hill Dr.
Knoxville, TN 37902
Phone: (615) 632-2101

In Washington:
412 1st St. SE, Ste. 300
Washington, D.C. 20444
Phone: (202) 479-4412
Fax: (202) 479-4421

U.S. Council For Energy Awareness (USCEA)
1776 Eye St. NW, Ste. 400
Washington, D.C. 20006
Phone: (202) 293-0770
Fax: (202) 785-4019

United Distribution Companies (UDC)
1225 W. 55th St.
New York, NY 10019
Phone: (212) 424-8120
Fax: (212) 424-3500

United States Energy Association (USEA)

1620 Eye St. NW, Ste. 210
Washington, D.C. 20006
Phone: (202) 331-0415
Fax: (202) 331-0418

Utilities Telecommunications Commission (UTC)

1140 Connecticut Ave. NW, Ste. 1140
Washington, D.C. 20036
Phone: (202) 872-1240
Fax: (202) 872-1331

Western Area Power Administration (WAPA)

P.O. Box 3402
Golden, CO 80401
Phone: (303) 231-1513
Fax: (303) 231-1632

In Washington:
1000 Independence Ave. SW, Ste. 8G061
Washington, D.C. 20585
Phone: (202) 586-5581
Fax: (202) 586-6261

Power Marketers And Contacts

(Author Note: Companies owned by electric and/or gas utilities or their holding companies are designated with an asterisk. These are emphasized due to their relatively new entrance into power and energy marketing activity.)

ACME Power Marketing, Inc.
c/o Dickstein, Shapiro & Morin
2101 L. St. NW
Washington, D.C. 20037
Contact: Patrick W. Lynch
Phone: (202) 785-9700
Fax: (202) 887-0689

AES Power, Inc.
1904 Plum Hill Court
Floyds Knobs, IN 47119
Contact: Bill Dukes
Phone: (812) 948-8442
Fax: (812) 948-8346

AIG Trading Corporation
One Greenwich Plaza
Greenwich, CT 06830
Contact: Michael A. Goldstein
Phone: (203) 861-3343
Fax: (203) 861-3822

Alliance Strategies, Inc.
18861 SW Martinazzi Ave., Ste. 203 E
Tualatin, OR 97062
Contact: J. Tim Watson
Phone: (503) 691-8577
Fax: (503) 691-9690

American National Power, Inc.
Boot Mills South
Foot of John Street
Lowell, MA 01852
Contact: Robert J. Kasle
Phone: (508) 452-2749
Fax: (508) 458-4287

American Power Exchange Corp.
9040 Executive Park Drive, Ste. 315
Knoxville, TN 37923
Contact: Eric Patterson
Phone: (615) 694-8222
Fax: (615) 693-0299

American Power Exchange, Inc.
598 S.W. Hidden River Ave.
Palm City, FL 34990
Contact: Richard A. Zambo
Phone: (407) 220-9163
Fax: (407) 220-9402

AMOCO Power Marketing Corp.
550 Westlake Park Blvd.
Houston, TX 77253
Contact: Tom Saile
Phone: (713) 366-7530
Fax: (713) 366-5901

ANP Energy Direct Company
10000 Memorial Drive, Ste. 500
Houston, TX 77024
Contact: Elaine G. Rollins
Phone: (713) 613-4363

Aquila Power Corporation*
10700 E. 350 Highway
Kansas City, MO 64138
Contact: Robert Parkey
Phone: (816) 737-7600
Fax: (816) 737-7985
**Owned by UtiliCorp United Inc.*

Ashton Energy Corporation
3100 Wilcrest Dr., Ste. 220
Houston, Texas 77042
Corporate Contact: Calvin Suddeath
Phone: (713) 781-3402
Fax: (713) 781-6709

Associated Power Services, Inc.
6120 S. Yale, Ste 1200
Tulsa, OK 74136
Contact: George W. Drake
Phone: (918) 492-3331
Fax: (918) 492-3375

Audit Pro, Inc.
World Trade Center, Ste. 114
San Francisco, CA 94111
Contact: Thomas Ringelmann
Phone: (415) 678-2960
Fax: (415) 296-7476

Bonneville Fuels Management Corp.
50 West 300 South, Ste. 600
Salt Lake City, UT 84101
Contact: Steve Stepanek
Phone: (801) 363-2520
Fax: (801) 363-9557

Boyd, Rosene & Associates, Inc.

1200 Park Centre, 525 S. Main
Tulsa, OK 74103
Contact: Michael K. Boyd
Phone: (918) 583-8617
Fax: (918) 582-4143

Btu Power, Inc.

777 108th Ave., NE, Ste. 1750
Bellevue, WA 98004
Contact: Robert E. Divers
Phone: (206) 455-4499

C.C. Pace Energy Services

4375 Fair Lakes Court, Ste. 2000
Fairfax, VA 22033
Contact: Larry Freedman
Phone: (703) 631-6600
Fax: (703) 378-1589

Calpine Power Marketing

50 W. San Fernando, 5th Floor
San Jose, CA 95113
Contact: Katherine Potter
Phone: (408) 995-5115
Fax: (408) 595-0505

Catex Vitol Electric, Inc.

470 Atlantic Ave., 12th Floor
Boston, MA 02210
Contact: Michael Kutsch
Phone: (617) 426-0777
Fax: (617) 426-3151

Cenerprise, Inc.*

1 Main Street SE, Ste. 600
Minneapolis, MN 55414
Contact: Thomas J. Lewis
Phone: (612) 331-4646
Fax: (612) 331-4641
**Owned by Northern States Power Co.*

Chicago Energy Exchange
205 N. Michigan Ave., Ste. 4315
Chicago, IL 60601
Contact: Lewis Babinski
Phone: (312) 819-2222
Fax: (312) 819-2221

Citizens Lehman, L.P.
530 Atlantic Ave.
Boston, MA 02210
Contact: Paul Barber
Phone: (617) 338-6300
Fax: (617) 951-0490

Citizens Lehman Power, L.P.
530 Atlantic Ave.
Boston, MA 02210
Contact: William E. James
Phone: (617) 338-6300
Fax: (617) 772-7550

CLP Hartford Sales, L.L.C.
530 Atlantic Ave.
Boston, MA 02210
Contact: William E. James
Phone: (617) 338-6300
Fax: (617) 772-7550

CMEX Energy, Inc.
17101 Preston Rd., Ste. 240
Dallas, Texas 75248
Contact: Norm Berthusen
Phone: (214) 248-2879
Fax: (214) 248-3553

CNB/Olympic Gas Services
14 E. University Ave., Ste. 213
Gainesville, FL 32601
Contact: Darin Cook
Phone: (904) 336-0202
Fax: (904) 336-7390

CNG Energy Services Corp.
P.O. Box 15746
Pittsburgh, PA 15244
Contact: Thomas E. Dodd
Phone: (412) 787-4286
Fax: (412) 787-6159

CNG Power Services
P.O. Box 15746
Pittsburgh, PA 15244
Contact: Dennis Lawler
Phone: (412) 787-4000
Fax: (412) 787-4260

Coastal Electric Services Co.
9 Greenway Plaza, Ste. 2919
Houston, Texas 77046
Contact: David W. Carey
Phone: (713) 877-7093
Fax: (713) 297-1184

CoEnergy Trading Co.
150 W. Jefferson, Ste. 1800
Detroit, MI 48226
Contact: Glen D. Kinder
Phone: (313) 256-5629
Fax: (313) 256-5739

Cogentrix Energy, Inc.
9405 Arrowpoint Blvd
Charlotte, NC 28273
Contact: Dennis W. Alexander
Phone: (704) 525-3800
Fax: (704) 529-1006

ConAgra Energy Services Co., Inc.
Nine Conagra Dr.
Omaha, NE 68102
Contact: Corey Amthor
Phone: (402) 595-4755
Fax: (402) 595-4759

CONOCO Power Marketing, Inc.
600 N. Dairy Ashford
Houston, TX 77079
Contact: Gordon E. Goodman
Phone: (713) 293-5152
Fax: (713) 293-5019

Coral Power, L.L.C.
909 Fanin, Ste. 700
Houston, TX 77010
Contact: Joe Bob Perkins
Phone: (713) 767-5409
Fax: (713) 767-5440

CRSS Power Marketing
1177 West Loop South
Houston, Texas 77027
Contact: Paul J. Margaritis
Phone: (713) 599-2656
Fax: (713) 599-2020

Dartmouth Power Associates, L.P.
One Energy Rd.
Dartmouth, MA 02747
Contact: Mitchell Jacobs
Phone: (508) 998-8515
Fax: (508) 998-7478

DC Tie
c/o Ritts, Brickfield & Kaufman
600 Watergate Bldg., Ste. 915
Washington, D.C. 20037
Phone: (407) 220-4480

Delhi Energy Services, Inc.
1700 Pacific Av., LB-10
Dallas, TX 75201
Contact: John E. Moeller, Jr.
Phone: (214) 954-2031
Fax: (214) 954-2700

Destec Energy Services Co.
2500 CityWest Blvd., Ste. 150
Houston, Texas 77042
Contact: Eric Dennison
Phone: (713) 735-4595
Fax: (713) 735-4678

Direct Access Management, L.P.
Box 204
Salsbury, VT 05769
Contact: L. Arthur Doty
Phone: (802) 352-4344
Fax: (802) 388-0450

Direct Electric, Inc.
140 Beachway Ave.
Palm City, FL 34990
Contact: Alan K. Richards
Phone: (407) 220-4480
Fax: (407) 220-1869

Duke Energy Marketing Corp.*
400 S. Tryon St., Ste. 1800
Charlotte, NC 28285
Contact: M. Rhem Wooten
Phone: (704) 382-2520
Fax: (704) 382-9325
**Owned by Duke Power Co.*

Duke/Louis Dreyfus Energy Services
New Englan, L.L.C.
10 Westport Rd.
Wilton, CT 06987
Contact: Ruby H. Melton, Esq.
Phone: (203) 761-2068

Duke/Louis Dreyfus, L.L.C.*

10 Westport Rd.
Wilton, CT 06897
Contact: Paul Addis
Phone: (203) 761-8141
Fax: (203) 761-8173
**Owned in part by Duke Energy Corp. and Duke Energy Marketing Corp.*

E Prime*

1331 17th St., Ste. 201
Denver, CO 80202
Contact: Mike Quigley
Phone: (303) 296-3513
Fax: (303) 296-3659
**Owned by Public Service Company of Colorado*

Eastern Power Distribution, Inc.

2900 Eisenhower Ave., Ste. 300
Alexandria, VA 22314
Contact: Ralph Bradley
Phone: (703) 317-2300
Fax: (703) 317-2301

Eastex Power Marketing, Inc.

1000 Louisiana St., Ste. 1100
Houston, TX 77002
Contact: Don B. Whaley
Phone: (713) 653-5469
Fax: (713) 650-6703

Eclipse Energy, Inc.

333 N. Sam Houston Pkwy E, Ste. 610
Houston, Texas 77060
Contact: J.Q. Delap
Phone: (713) 999-8200
Fax: (713) 447-1009

EDC Power Marketing, Inc.
1817 Wood St., Ste. 550 West
Dallas, Texas 75201
Contact: Peggy J. Banczak, Esq.
Phone: (214) 573-3084
Fax: (214) 670-2974

Electrade Corp.
200 Liberty St., 23rd Floor, Tower A
New York, NY 10281
Contact: Preston D. Head
Phone: (212) 693-2020
Fax: (212) 565-5514

Electric Clearinghouse, Inc.
13430 Northwest Freeway, Ste. 1200
Houston, Texas 77040
Contact: Daniel Ryser
Phone: (713)507-6500
Fax: (713) 507-6505

Energy Alliance Partnership
One Park Ridge Center
P.O. Box 15746
Pittsburgh, PA 15244
Contact: Holly Koeppel
Phone: (412) 787-4278
Fax: (412) 787-4464

Energy Marketing Services, Inc.
355 E. Campus View Blvd., Ste. 150
Columbus, OH 43235
Contact: Jeffrey A. Jones
Phone: (614) 431-1858
Fax: (614) 431-1862

EnergyOnline, Inc.
4962 El Camino Real, Ste. 112
Los Altos, CA 94022
Contact: Richard Albert
Phone: (415) 962-9670
Fax: (415) 962-9615

Energy Resource Management Corp.
P.O. Box 44563
Detroit, MI 48244
Contact: Ronnie James
Phone: (313) 965-2721
Fax: (313) 965-2614

Energy Resource Marketing, Inc.
800 Country Place Drive
Houston, TX 77079
Contact: MaryJane Reynolds
Phone: (202) 331-8000

Energy Services, Inc.
101 E. 8th St., Ste. 240
Vancouver, WA 98660
Contact: Noel Shelton
Phone: (360) 694-2894
Fax: (360) 694-3082

Energy Transfer Group, L.L.C.
8080 North Central Expressway, Ste. 650
Dallas, TX 75206
Contact: Cliff Harris
Phone: (214) 368-6530
Fax: (214) 368-2112

Energy West Power Co.
5674 Stoneridge Dr., Ste. 108
Pleasanton, CA 94588
Contact: John Tomasello
Phone: (510) 847-8910
Fax: (510) 847-8915

Enerserve, L.C.

P.O. Box 54618
Tulsa, OK 74155
Contact: Jeffery A. Milford
Phone: (918) 369-6977
Fax: (918) 494-7996

Engelhard Power Marketing, Inc.

101Wood Ave.
Iselin, NY 08830
Contact: Joseph Racelis
Phone: (908) 205-6364
Fax: (908) 205-6290

Enova Energy Management, Inc.*

P.O. Box 1831
San Diego, CA 92112
Contact: James F. Walsh, III
Phone: (619) 699-5022
Fax: (619) 699-5027
**Owned by San Diego Gas & Electric Co.*

Enpower, Inc.

3903 Turkey Oak Dr.
Valrico, FL 33594
Contact: Daniel Smith
Phone: (813) 681-8775
Fax: (813) 681-3227

Enron Power Marketing, Inc.

P.O. Box 1188
Houston, TX 77251
Contact: Steven J. Kean
Phone: (713) 853-1586
Fax: (713) 646-8160

Entergy Power Marketing Corp.*

900 S. Shackleford Rd.
Little Rock, AR 72211
Contact: Frederick. F. Nugent, Esq.
Phone: (501) 954-5000
Fax: (501) 954-5003
**Owned by Entergy Corp.*

Equitable Power Services Company

200 West Lake Park Blvd.
Houston, TX 77079
Contact: Boyd L. Henderson
Phone: (713) 560-7808
Fax: (713) 560-7890

Euroamerican Energy Services

1980 Post Oak Blvd., St. 2010
Houston, TX 77056
Contact: Michael Moore
Phone: (713) 850-9866
Fax: (713) 850-7691

Eurobrokers

2 World Trade Center,Ste. 8400
New York, NY 10048
Contact: Jay Perl
Phone: (212) 748-7250
Fax: (212) 869-4521

Excel Energy Services, Inc.

37543 E. Greenwood
Northville, MI 48167
Contact: Paul S. Bahra
Phone: (810) 471-2111
Fax: (810) 442-7820

Federal Energy Sales, Inc.
3222 N. Ridge Rd. East
Elyria, OH 44035
Contact: Scott S. Towner
Phone: (216) 324-9950

Gateway Energy Marketing
300 N. Lake Ave., Ste. 520
Pasadena, CA 91101
Contact: Art. M. Gastelum

GatewayEnergy, Inc.
3556 Lakeshore Rd., Ste. 730
Buffalo, NY 14219
Contact: Mark V. Cuda
Phone: (716) 824-2909
Fax: (716) 824-3309

GED Gas Services, L.L.C.
7666 E. 61st St., Ste. 370
Tulsa, OK 74133
Contact: Bob Jones
Phone: (918) 254-1531
Fax: (918) 254-6903

Global Petroleum Corp.
P.O. Box 9161
Waltham, MA 02254
Contact: Edward J. Faneuil, Esq.
Phone: (617) 398-4211
Fax: (617) 398-4160

Greenwich Energy Partners, L.P.
140 Greenwich Ave.
Greenwich, CT 06830
Contact: Lance A. Bakrow
Phone: (203) 861-7809
Fax: (203) 861-6603

Gulfstream Energy, LLC

2930 Revere St., Ste. 202
Houston, TX 77098
Contact: Frank Spann
Phone: (713) 527-0019
Fax: (713) 527-0828

Hadson Electric, Inc.

2777 Stemmons Fwy., Ste. 700
Dallas, TX 75207
Contact: Paul Brundage
Phone: (214) 640-6800
Fax: (214) 640-6933

Heartland Energy Services, Inc.*

802 W. Broadway, Ste. 301
Madison, WI 53713
Contact: Richard Friedman
Phone: (608) 221-7700
Fax: (608) 221-7710
**Owned by WPL Holdings Inc.*

Heath Petra Resources, Inc.

5599 San Felipe, Ste. 580
Houston, TX 77056
Contact: Douglas Morgan
Phone: (713) 961-1281
Fax: (713) 961-3862

Hinson Power Co.

40 Lake Bellevue, Ste. 100
Bellevue, WA 98005
Contact: James D. Stromberg
Phone: (206) 450-4010
Fax: (206) 450-5569

Howard Energy Company, Inc.
7502 W. Oakhill Ave.
Wauwatosa, WI 53213
Contact: John P. Raisler
Phone: (414) 774-0155
Fax: (414) 774-0156

Howell Power Systems, Inc.
1010 Lamar
Houston, TX 77002
Contact: Gordon Goodman
Phone: (713) 658-4100
Fax: (713) 658-4088

ICPM, Inc.
1130 Lake Cook Rd., Ste. 300
Buffalo Grove, IL 60089
Contact: John W. Salyer, Jr.
Phone: (708) 520-3212
Fax: (708) 520-9883

IEP Power Marketing, L.L.C.
12150 E. Briarwood Ave., Ste. 145
Englewood, CO 80112
Contact: Alan K. Forbes

IGI Resources, Inc.
300 Mallard Dr., Ste. 350
Boise, ID 83706
Contact: Randy Schultz
Phone: (208) 387-0500
Fax: (208) 387-0530

IGM, Inc.
1635 W. First Avenue
Columbus, OH 43212
Contact: Ronald B. Williamson
Phone: (614) 487-4595
Fax: (614) 481-3482

Illinova Power Marketing, Inc.*

1405 W. 2200 South, Ste. 201
Salt Lake City, UT 84119
Contact: Robert A. Schultz
Phone: (801) 978-8150
Fax: (801) 973-8924
**Owned by Illinova Power Corp.*

Imprimis Corporation

508 Drake Ave.
Centerville, IA 52544
Contact: Thomas L. Anders
Phone: (515) 437-1422
Fax: (515) 437-1970

Industrial Energy Applications*

P.O. Box 351
Cedar Rapids, IA 52406
Contact: John Bundy
Phone: (319) 378-6508
Fax: (319) 378-6565
**Owned by IES Industries Inc.*

Industrial Gas and Electric Services

5301 Central Avenue, N.E., Ste. 1213
Albuquerque, NM 87108
Contact: Alice Robledo
Phone: (505) 265-3778
Fax: (505) 265-3942

InterCoast Energy Company

P.O. Box 657
Des Moinest, IA 50303
Contact: Bruce E. Weier
Phone: (515) 281-2671
Fax: (515) 281-2312

International Energy Co.
37155 Golfview Dr.
Sterling Heights, MI 48312
Contact: W. O'Brian Davis
Phone: (810) 939-3461
Fax: (810) 939-3461

J Power
P.O. Box 774
Wayzata, MN 55391
Contact: John S. Jaffray
Phone: (612) 476-8740
Fax: (612) 476-9383

J. Anthony & Associates, Ltd.
3986 St. Mary's Avenue
N. Vancouver, BC V7N 1Y3
Canada
Contact: John Anthony
Phone: (604) 987-2775
Fax: (604) 987-6017

J. Aron/Goldman Sachs
85 Broad St., 5th Floor
New York, NY 10004
Contact: Thadeus Miller, Esq.
Phone: (212) 902-5394
Fax: (212) 902-3876

J.D. Loock & Associates
5383 Mariner's Cove Dr., Ste. 411
Madison, WI 53704
Contact: James D. Loock
Phone: (608) 242-1194

JEB Corporation
3420 Miramar Dr.
Pine Bluff, AR 71603
Contact: James E. Brabston, Jr.
Phone: (501) 534-5905

Jefferson Electric Marketing, Inc.
1530 Wilson Blvd., Ste. 900
Arlington, VA 22209
Contact: Vinod K. Dar
Phone: (703) 351-0300
Fax: (703) 351-0352

J.L. Walker & Associates
7434 Cedear Creek Trail
Madison, WI 53717
Contact: John L. Walker
Phone: (608) 829-0538
Fax: (608) 829-0550

K Power Company, Inc.
80 Eighth Avenue, Ste. 1107
New York, NY 10011
Contact: Philip Kielkucki
Phone: (212) 206-7414
Fax: (212) 206-1534

Kaztex Energy Services, Inc.
350 Bishops Way, P.O. Box 1040
Brookfield, WI 53008
Contact: John R. Krieger
Phone: (414) 784-1133
Fax: (414) 784-2925

KCS Energy Management Service, Inc.
379 Thornall St.
Edison, NJ 08837
Contact: Tom Withka
Phone: (908) 632-1770
Fax: (908) 494-3253

KCS Energy Marketing, Inc.
1800 West Loop South, Ste. 1400
Houston, TX 77027
Contact: Harry Lee Stout
Phone: (713) 877-8006
Fax: (713) 877-1394

Kimball Power, Inc.

32270 Telegraph Rd., Ste. 200
Birmingham, AL 48025
Contact: James A. Nichols
Phone: (810) 642-6509
Fax: (810) 647-9735

KinEr-G Power Marketing, Inc.

c/o Metzger, Hoolis, Gordon & Alprin
1275 K Street NW, Ste. 1000
Washington, D.C. 20005
Contact: Carol A. Smoots, Esq.
Phone: (202) 842-1600
Fax: (202) 682-2127

KN Marketing, Inc.

P.O. Box 281304
Lakewood, CO 80228
Contact: S. Wesley Haun
Phone: (303) 763-3420
Fax: (303) 763-3511

Koch Power Services, Inc.

4111 E. 37th St. N.
Wichita, KS 67220
Contact: Stefan A. Bird
Phone: (316) 828-5245
Fax: (316) 828-7977

Kohler Company

444 Highland Dr.
Kohler, WI 53044
Contact: Michael J. Potts
Phone: (414) 457-4441
Fax: (414) 457-1663

LG&E Power Marketing, Inc.*
12500 Fair Lakes Circle, Ste. 350
Fairfax, VA 22033
Contact: B. Jeanine Hull
Phone: (703) 968-7200
Fax: (703) 968-5458
**Owned by LG&E*

Louis Dreyfus Electric Power, Inc.
Ten Westport Rd.
Wilton, CT 06897
Contact: Paul Addis
Phone: (203) 761-2116
Fax: (203) 761-8173

Market Responsive Energy, Inc.*
6200 Oak Tree Blvd. IND-330
Independence, OH 44256
Contact: Richard R. Holmes, Jr.
Phone: (216) 447-2755
Fax: (216) 447-3599
**Owned by Centerior Energy Corp.*

Mesquite Energy Services, Inc.
3845 FM 1960 West, Ste. 250
Houston, TX 77068
Contact: Sam C. Henry
Phone: (713) 893-2050
Fax: (713) 379-6750

MG Electric Power, Inc.
1000 Louisiana, Ste. 6600
Houston, TX 77002
Contact: Erik W. Herbst
Phone: (713) 756-7708
Fax: (713) 659-8510

Mid-American Resources, Inc.*
4630 Arcady Ave.
Dallas, TX 75209
Contact: J. Todd Moore
Phone: (214) 523-9089
Fax: (214) 523-9001
**Owned by IPALCO Enterprises Inc.*

MidCon Power Services Corp.
701 E. 22nd St.
Lombard, IL 60148
Contact: Fred Jahns
Phone: (708) 691-2539
Fax: (708) 691-3827

Morgan Stanley Capital Group
1221 Avenue of the Americas
New York, NY 10020
Contact: Neil Levy
Phone: (212) 296-2554
Fax: (212) 296-2485

MP Energy, Inc.
40 East Broadway
Butte, MT 59701
Contact: Richard F. Cromer
Phone: (406) 486-5388
Fax: (406) 496-5396

MPEX*
30 West Superior St.
Duluth, Minn. 55802
Contact: Mike Critchley
Phone: (218) 720-2770
**Owned by Minnesota Power*

Multi-Energies USA, Inc.
1 Commerce Place, Ste. 500
Verdun, PQ H3E 1A2
Canada
Contact: Pierre Gravel
Phone: (514) 769-6048
Fax: (514) 766-4236

NAP Trading and Marketing, Inc.
8480 E. Orchard Rd., Ste. 4000
Englewood, CO 80111
Contact: Michael J. Ruffatto
Phone: (303) 796-8600
Fax: (303) 773-0461

National Electric Associates, L.P.
3555 Timmons, Ste. 1500
Houston, TX 77027
Contact: Marjorie N. Glasgow
Phone: (713) 961-1890
Fax: (173) 961-1153

National Fuel Resources, Inc.
478 Main St.
Buffalo, NY 14202
Contact: Stephen J. Muscato
Phone: (716) 857-6732
Fax: (716) 857-6798

National Power Exchange Corp.
P.O. Box 4455
Houston, TX 77210
Contact: Michael L. Wallace
Phone: (713) 654-5600
Fax: (713) 654-5977

National Power Management Co.
P.O. Box 7107
McLean, VA 22106-7107
Contact: Glenn Williams
Phone: (703) 356-8333
Fax: (703) 847-9191

NFR Power, Inc.
478 Main St.
Buffalo, NY 14202
Contact: Robert J. Kreppel
Phone: (716) 857-6791
Fax: (716) 857-6798

NorAm Energy Services, Inc.
P.O. Box 4455
Houston, TX 77210
Contact: Michael L. Wallace
Phone: (713) 654-5959
Fax: (713) 654-5513

Nordic Electric, L.L.C.
2010 Hogback Rd., Ste. 4
Ann Arbor, MI 48105
Contact: Walter J. Balcer
Phone: (313) 973-7700
Fax: (313) 973-9335

Norstar Energy, L.P.
28 Grand Avenue
Montvale, NY 07645
Contact: Nicholas W. Mattia, Jr.
Phone: (201) 307-1188

North American Energy Corp.
280 Park Avenue, Ste. 2700 West
New York, NY 10017
Contact: Robert M. Beningson
Phone: (212) 557-6200
Fax: (212) 557-5678

North American Power Brokers, Inc.
52 October Hill Rd.
Hamden, CT 06518
Contact: John P. Gaus
Phone: (203) 281-7132

Northwest Power Marketing Co.
1201 Walnut
Kansas City, MO 64106
Contact: Mark G. English
Phone: (816) 556-2784
Fax: (816) 556-2802

Ocean Energy Services, Inc.
174 Ocean Ave., Ste. 2
Sea Bright, NJ 07760
Contact: George R. Henderson
Phone: (908) 741-6455
Fax: (908) 741-4643

Oxbow Power Marketing, Inc.
1601 Forum Place, Ste. P-2
West Palm Beach, FL 33401
Contact: David W. Clark, Esq.
Phone: (407) 697-4300

PacifiCorp Power Marketing, Inc.*
825 N.E. Multnomah St., Ste. 485
Portland, OR 97232
Contact: Donald N. Furman
Phone: (503) 464-5071
Fax: (503) 275-2714
**Owned by PacifiCorp*

PanEnergy Power Service, Inc.
P.O. Box 1642
Houston, TX 77215
Contact: Doug Wilner
Phone: (713) 627-5085
Fax: (713) 989-1588

Paragon Gas Marketing
1000 Louisiana, Ste. 1100
Houston, TX 77002
Contact: Randy Haws
Phone: (713) 961-1100
Fax: (713) 961-0555

PennUnion Energy Services, L.L.C.
1330 Post Oak Blvd., Ste. 2000
Houston, TX 77056
Contact: Trisha Pollard, Esq.
Phone: (713) 297-9543
Fax: (713) 621-5392

Peak Energy, Inc.
50 W. Broadway, Ste. 1111 Bank One Tower
Salt Lake City, UT 84101
Contact: Jeff J. Fishman
Phone: (801) 596-0500
Fax: (801) 596-3434

Petroleum Source & Systems Group, Inc.
P.O. Box 900038
Atlanta, GA 30329
Contact: Livia Whisenhunt
Phone: (404) 321-5711
Fax: (404) 321-3938

Philbro, Inc.
500 Nyala Farm Rd.
Westport, CT 06880
Contact: Thomas Barry
Phone: (203) 221-5910
Fax: (203) 221-6772

Power Clearinghouse
12121 N.W. Freeway, Ste. 130
Houston, TX 77092
Contact: Joelle S. Barnes
Phone: (713) 460-2200
Fax: (713) 460-8611

Power Exchange Corporation
500 Quail St., Ste. 550
Newport Beach, CA 92660
Contact: William C. Prentice
Phone: (714) 263-5929
Fax: (714) 261-2824

Powerex
666 Burrard St.
Vancouver, BC V6C2X8
Contact: Zak El-Ramly
Phone: (604) 891-5005
Fax: (604) 891-5015

PowerMark, L.L.C.
P.O. Box 6303
Denver, CO 80206
Contact: Raymond D. Danton
Phone: (303) 777-7236
Fax: (303) 777-7583

PowerNet Corp.
5301 N. Federal Highway, Ste. 130
Boca Raton, FL 33487
Contact: Anthony F. Altmann
Phone: (407) 998-8500
Fax: (407) 998-8380

Power Smart, Inc.
231 Salina Meadows Pkwy., Ste. 185
P.O. Box 4743
Syracuse, NY 13221-4743
Contact: John W. Wilson
Phone: (315) 453-7442
Fax: (315) 453-0067

Powertec International, L.L.P.
P.O. Box 1716
Smithfield, NC 27577
Contact: S. Barry Hales
Phone: (919) 934-5115
Fax: (919) 934-5118

Prairie Winds Energy, Inc.
P.O. Box 1791
Bismarck, ND 58502
Contact: Ronald A. Hodge
Phone: (701) 224-0626
Fax: (701) 224-0017

Premier Enterprises, Inc.
5670 Greenwood Plaza Blvd., Ste. 420
Englewood, CO 80111
Contact: Richard C. Frantz
Phone: (303) 694-3836
Fax: (303) 694-3835

Progas Power, Inc.
201 E. Gibson St.
P.O. Box 0974
Covington, LA 70434
Contact: John S. Burge
Phone: (504) 893-5833
Fax: (504) 893-7697

Proven Alternatives, Inc.
1740 Army St.
San Francisco, CA 94124
Contact: Vernon A. Harris
Phone: (415) 285-0800
Fax: (415) 285-5805

QST Energy Trading, Inc.
300 Hamilton Blvd., Ste. 330
Peoria, IL 61602
Contact: Keith Goerss
Phone: (309) 655-1355
Fax: (309) 655-1366

Questar Energy Trading Co.
P.O. Box 45433
Salt Lake City, UT 84145
Contact: Tom Jepperson
Phone: (801) 530-2648
Fax: (801) 530-2637

Rainbow Energy Marketing Corporation
919 S. Seventh St., Ste. 405
Bismark N.D. 58504
Contact: Loren R. Kopseng
Phone: (701) 255-7970
Fax: (701) 255-7952

Rig Gas, Inc.
500 N. Third St., Ste. 109
Fairfield, IA 52556
Contact: Robert Markowitz
Phone: (515) 469-6666
Fax: (515) 472-5540

R.J. Dahnke & Associates
P.O. Box 1254
Sheboygan, WI 53082
Contact: Richard J. Dahnke
Phone: (414) 457-5070
Fax: (414) 457-5134

Ruffin Energy Services, Inc.
5530 SW Highway 66
Claremore, OK 74017
Contact: Roger P. Gleason
Phone: (918) 266-4100
Fax: (918) 266-4125

SCANA Energy Marketing, Inc.*
7909 Parklane Rd.
Columbia, SC 29223
Contact: Charles A. Rampey, Jr., P.E.
Phone: (803) 741-1052
Fax: (803) 741-1058
**SCANA Corporation*

Seagull Power Services, Inc.
1001 Fannin, Ste. 1700
Houston, TX 77002
Contact: John R. Medler
Phone: (713) 951-4747
Fax: (713) 951-4819

Sonat Power Marketing, Inc.
P.O. Box 2563
Birmingham, AL 35202
Contact: Linda K. Browning
Phone: (205) 325-7599
Fax: (205) 325-3850

Southeastern Energy Resources, Inc.
1565 Woodington Circle, Ste. 202
Lawrenceville, GA 30244
Contact: Paul Hyland
Phone: (770) 806-9432
Fax: (770) 806-9482

Southern Energy Marketing, Inc.
900 Ashwood Parkway, Ste. 500
Atlanta, GA 30338
Contact: Marce Fuller
Phone: (770) 379-7647
Fax: (770) 379-7718

Stalwart Power Co.
1 West 3rd St., Ste. 1320
Williams Center, Tower 1
Tulsa, OK 74103
Contact: Debbie Branch
Phone: (918) 584-7799
Fax: (918) 584-5771

Stand Energy Corporation
1077 Celestial St./Rookwood Bldg., Ste. 10
Cincinnati, OH 45202
Contact: Robert P. Robison
Phone: (513) 621-1113
Fax: (513) 621-3773

SuperSystems, Inc.
17561 Teachers Ave.
Irvine,CA 92714
Contact: Sam Tadros
Phone: (714) 786-7117
Fax: (714) 733-3430

Superior Electric Power Corp.
1021 Main St., Ste. 2100
Houston, TX 77002
Contact: John W. Croft, Jr.
Phone: (713) 759-6900
Fax: (713) 756-1133

Tenaska Power Services Co.
1044 N. 115th St., Ste. 400
Omaha, NE 68154
Contact: Larry Pearson
Phone: (402) 691-9505
Fax: (402) 691-9538

Tenneco Energy Marketing Co.
P.O. Box 2511
Houston, TX 77252
Contact: Randall M. Thompson
Phone: (713) 757-5876
Fax: (713) 757-4402

Tennesssee Power Company
4612 Maria St.
Chattanooga, TN 37411
Contact: Michael R. Knauff, P.E.
Phone: (615) 624-0852
Fax: (615) 624-4906

Texaco, Inc.
1111 Bagby, Ste. 2788
Houston, TX 2788
Contact: John P. Beall
Phone: (713) 752-6001
Fax: (713) 752-4762

Texas-Ohio Power Marketing, Inc.
One Memorial City Plaza, 800 Gessner, Ste. 900
Houston, TX 77024
Contact: T. Pat Harrison
Phone: (713) 468-4686
Fax: (713) 468-1457

Texican Energy Ventures, Inc.
505 N. 20th St.
Birmingham, AL 35203
Contact: Kevin Stump
Phone: (205) 251-8081
Fax: (205) 251-8084

TexPar Energy, Inc.
2020 Springdale Rd.
Waukesha, WI 53186
Contact: Robert Rivas
Phone: (414) 548-1488
Fax: (414) 548-1483

Torco Energy Marketing, Inc.
111 E. Wacker Dr., Ste. 1300
Chicago, IL 60601
Contact: Michael Prior
Phone: (312) 616-1700
Fax: (312) 616-9388

TransAlta Corporation
110-12 Ave. SW
Calgary, AB T2P 2M1
Canada
Contact: Terence Dalgleish, Q.C.
Phone: (403) 267-4761
Fax: (403) 267-3630

TransCanada Northridge Power, Ltd.
421 7th Ave. SW, Ste. 1400
Calgary, AB T2P 4K9
Canada
Contact: Mark Himmelspach
Phone: (403) 262-6800
Fax: (403) 262-0113

Transco Power Trading Co.
Memorial City Place, 9821 Katy Freeway
Houston, TX 77024
Contact: James P. Sale
Phone: (713) 932-4845
Fax: (713) 932-4120

Universal Power Services, L.L.C.
12301 Stoney Creek Rd.
Potomac, MD 20854
Contact: Dean C. Lovett
Phone: (301) 926-9360

U.S. Power & Light, Inc.
10107 Chevy Chase
Houston, TX 77042
Contact: Charles Manning
Phone: (713) 782-8037

Utility Management & Consulting
15609 North 55th St.
Scottsdale, AZ 85254
Contact: William B. McNally, III
Phone: (602) 867-8099
Fax: (602) 867-1861

Utility Trade Corp.
140 4th Ave. SW, Ste. 1710
Calgary, AB T2P 3N3
Canada
Contact: Darcy J.C. White
Phone: (403) 531-2699
Fax: (403) 531-2695

Utility-2000 Energy Corp.
140-4th Ave. S.W., Ste. 1710
Calgary, AB T2P 3N3
Canada
Contact: Darcy White
Phone: (403) 531-2699
Fax: (403) 531-2695

Valero Power Services Company
1200 Smith St., Ste. 900
Houston, TX 77002
Contact: Harry E. Watson
Phone: (713) 951-2262
Fax: (713) 951-2275

Vanpower, Inc.
P.O. Box 9805
Vancouver, WA 98660
Contact: Richard L. Smith
Phone: (360) 696-8661
Fax: (360) 696-8775

Vantus Energy Corp.*
444 Market St., Ste. 1900
San Francisco, CA 94111
Contact: Dale A. Murdock
Phone: (415) 291-6420
Fax: (415) 291-6473
**Owned by Pacific Gas & Electric Co.*

Vastar Power Marketing, Inc.
200 Westlake Park Blvd., Ste. 200
Houston, TX 77079
Contact: J.P. Catasein
Phone: (713) 584-3940
Fax: (713) 584-3901

Vesta Energy Alternatives Co.
100 W. 5th St., 320 Oneok Plaza
Tulsa, OK 74103
Contact: Christopher J. Bernard
Phone: (918) 599-1529
Fax: (918) 584-1855

VTEC Energy, Inc.
212 Manida St.
Bronx, NY 10474
Contact: Neil Schultz
Phone: (718) 542-8248
Fax: (718) 542-8759

WestCoast Power Marketing, Inc.
421 7th Ave., SW, Ste. 1100
Calgary, AB T2P 4K9
Canada
Contact: Peter Leier
Phone: (403) 297-0346
Fax: (403) 262-3510

Westar Electric Marketing, Inc.
818 Kansas Ave.
P.O. Box 889
Topeka, KS 66601
Contact: Rita A. Sharpe
Phone: (913) 575-8344
Fax: (913) 575-1771

Western Gas Resource Power Marketing
12200 N. Pecos St., Ste. 230
Denver, CO 80234
Contact: Bill Sanderson
Phone: (303) 452-5603
Fax: (303) 252-6150

Wheeled Electric Power Co.
50 Charles Lindbergh Blvd., Ste. 400
Uniondale, NY 11553
Contact: Dr. John O'Brien
Phone: (516) 229-2280
Fax: (516) 229-2382

Wholesale Power Services, Inc.*
251 N. Illinois St., Ste. 1410
Indianapolis, IN 46204
Contact: Robert Dyer
Phone: (317) 488-3410
Fax: (317) 488-3405
**Owned by Cinergy Corp.*

Wickford Energy Marketing, L.C.
2323 S. Shepherd, Ste. 810
Houston, TX 77019
Contact: Mike Jones
Phone: (713) 942-0595
Fax: (713) 942-7915

Wickland Power Services
P.O. Box 13648
Sacramento, CA 95853
Contact: Ross A. Clark
Phone: (916) 978-2470
Fax: (916) 978-2449

Wicor Energy Services Inc.
306 N. Milwaukee St.
Milwaukee, WI 53202
Contact: Chuck Cummings
Phone: (414) 291-6661
Fax: (414) 226-4508

Williams Energy Services Company
One Williams Plaza, Ste. 19-2
Tulsa, OK 74141
Contact: Baron Dean Thrower
Phone: (918) 588-4844
Fax: (918) 561-6987

WPS Energy Services Inc.
WPS Power Development Inc.
P.O. Box 19001
Green Bay, WI 54307
Contact: William L. Bourbonnais
Phone: (414) 433-1573
Fax: (414) 433-5734

Yankee Energy Marketing Co.

599 Research Parkway
Meriden, CT 06450
Contact: Page D. Miller
Phone: (203) 639-4672
Fax: (203) 639-4677

A Power Marketer's Business Plan

Author's Note: The following power marketer strategic business plan was authored by four students in an executive MBA class on strategic marketing at Loyola College in Maryland. The author/students are:

Charles R. Walker Jr.

H. Joseph Fleischman, P.E.

Robert Driscoll

Michael D. Carlisle

Students were asked to develop a scenario for various industries for the year 2005 based on key trends now at work in the competitive environment. The idea was to demonstrate the power of "future" thinking and how it applies to strategic planning. Students were introduced to the difficult task of dealing with the vagueness and ambiguity in coping with change.

CONTENTS

VISION STATEMENT

Electric Warehouse Power Marketing (EWPM) is formed to serve as an Independent Power Marketer and Broker. In this function, Electric Warehouse must continue to strengthen strategic business alliances and develop end-user products and strategies that allow for homes, municipalities, and industrial clients to have freedom of choice for their electric power supplier.

Electric Warehouse's efforts will continue to provide the most reliable and cost-effective supply of electricity to all of its clients.

EXECUTIVE SUMMARY

Electric Warehouse is a privately held electric power producer, marketer, and broker. After identifying customers who need electrical power, we acquire and re-deploy energy from a variety of supplies. Electric Warehouse may generate the electricity from its own facilities or find other producers, marketers, or brokers to provide the power and wheel the power to the users. We procure electricity at a favorable price and deliver this product efficiently to the customer.

Electric Warehouse does not play from the sidelines solely as a broker, but actively participates directly in the marketplace by producing and buying needed supplies and providing them to our customers on demand. Market knowledge, strategic partnering, and state of the art equipment allow us to seize opportunities and provide cost affective electric power solutions.

MARKET SEGMENTATION

The need to segment and target accurately will determine the successful power marketers, brokers, and producers well into the 21st century. This fact is why Electric Warehouse has chosen to segment the electric power market according to customer classification. It is critical that each factor is used in assessment due to fact that electricity costs vary region-to-region, utility-to-utility, REA-to-REA (rural electric association), and according to varying customer needs and desires.

Electric Warehouse has segmented the electric business in the United States as follows: customer classification, present supplier, state, region, 1995 MW percent attributed to each classification, and service desired. Each of these segmentation classifications will help Electric Warehouse determine the customers in which they wish to target.

The methods used will be different than utility competition since they typically focus on their captive market—their franchise territory. However, competition will attempt to develop similar model segments to focus on to win the most lucrative customers. Given the present electricity market accounting for over $200 billion in annual revenues in the United States alone there exists a potential equal or greater than this with additional services and products. Of course Electric Warehouse will target different segments of the overall market with different positioning techniques as the needs vary customer-to-customer and classification-to-classification.

MARKET TARGETING

With a well focused and accurate market segmentation, Electric Warehouse will be able to choose its battles wisely and avoid high costs associated with winning utility-loyal customers. This approach will be executed by using the following five-step method.

- Step 1: Determine high price and center on targeting utilities with lofty rates such as New York City Lighting, Consolidated Electron, Commonwealth Energy, and METRO Energy, for example purposes only.

- Step 2: Align Electric Warehouse with producers with access to adequate transmission systems in states that are welcoming deregulation such as Rhode Island, Vermont, New Hampshire,

California, Massachusetts, and Illinois and be poised to move immediately upon passage of such legislation.

- Step 3: Understand the need to lure the most lucrative customers away from their present providers. As long distance deregulation moved forward, the major long distance carrier lost 30% of its customers. These happened to be the lowest margin customers. Electric Warehouse must win higher margin customers.

- Step 4: Work to win high volume/lower margin customers to gain market share and develop alliances with suppliers and consumers. This may seem contrary to the third step.

- Step 5: Provide better service, become a true partner, and understand the customer and their business.

This five-step approach targets high price utilities; suppliers of choice; lucrative customers; high-volume, low-margin customers; and targets Electric Warehouse's position in the eyes of the consumer.

Another target market will be that in which customers use Electric Warehouse as a negotiating tool with their present provider. The company will offer expert negotiators as agents in cases where consumers feel comfortable with their present provider but only want lower rates. We do not wish to be only a bargaining tool. We will continue to concentrate on our core business of brokering, marketing, and producing energy products which are mutually beneficial to consumers and Electric Warehouse.

MARKET POSITIONING

Electric Warehouse will initially position itself as a low cost provider to large-volume, low-margin wholesale customers in the 1996 electricity market. This approach will allow the company the luxury of honing its marketing and brokering skills and gaining market share as a preface to full scale retail wheeling/deregulation.

Electric Warehouse, in this role, will learn the details of complex transmission systems, will learn how independent system operations work, and learn to understand how to satisfy the needs of the new electricity consumer.

Initially, this approach is required since the local utilities were first to market, like Coke, and have been doing a fairly good job at providing electric service. Second, cogenerators and alternate energy products have been subsidized, once pronounced as the wave of the future, and subsidized by the DOE and others.

In one context, marketers and brokers are thought of as the financial guys that may not know electricity as well as the local utility, but for the right price would be the provider of choice.

In another context, E.I. Dupont deNemours Company, Inc. feels price will be the determining factor on which they choose energy suppliers. Dupont presently spends $1,200,000,000 annually on energy. (Unless otherwise noted, this information was taken from interviews with Robert Freedman, Director of Energy E.I. Dupont deNemours, Inc., April, 1996.)

However, the future is upon us with utility mergers and acquisitions making local utilities regional, and sometimes national. With recent legislation we are no longer leaving the franchise territory to only the utility anymore. Deregulation will prevail over the next few years and, with its present position, Electric Warehouse will get its name on the street as a quality provider of energy solutions. Given this, Electric Warehouse must understand that today's constraints are only that, since future deregulation and open access will open all transmission systems to utilities, marketers, brokers, and anyone. In the future we may look and see utilities that were vertically integrated become buyers of power marketers' products and services. This is Electric Warehouse's goal.

Given Electric Warehouse's attempt to gain name recognition, it must also position to secure a name as a reliable source of energy products and services. The company must develop such market perception so as to remove gaps from what is expected and what is perceived. In the future, Electric Warehouse will need to supply electricity to consumers who desire reliable energy first and cheap energy second as well as vice versa. All these requirements will allow Electric Warehouse to use its market leadership in marketing, brokering, logistics, and production and to provide products and services when they need to at a price that is beneficial to the consumer and to the company.

OUR PRODUCT

Our product is more than simply electrons pushed through copper wires. Power companies have done that for years. Our electricity product includes convenience, efficiency, and reliability. Our function appears simple at first: we purchase surplus power production during off-peak times and wheel that power to consumers.

This method, to the extent possible, benefits consumers with lower prices. Because we are also a producer of power, we are able to supplement our purchases with our own production.

Clearly, electricity is a high volume product. In effect it is a commodity. Electricity is a commodity upon which our society depends upon heavily for a constant supply. Society grinds to a halt during periods, such as bad weather, when electric service is interrupted.

This demand and dependence upon electric power has been built over the past. As the population moved west, new power production facilities grew. People demanded service and companies were happy to provide. Before the 1980s, every power company produced the same product for the market surrounding its production facilities. Customers knew only that they could flip the switch and our products turned on the television. They knew that they got a bill each month and knew that rates were regulated by the government.

Then talk of deregulation began to surface and consumer product preferences began to change. Technology began to allow customers to require a higher level of convenience from service providers. No longer did customers feel that they had to settle for status quo. Our advantage over other producers is that we have customized our billing and service products to be compatible with local television service. This allows a high degree of convenience and control to be available to our target household consumers and a higher degree of product recognition. Electric Warehouse was the first company to combine into a single product all three aspects of the power production and supply industry: producer, broker, and marketer.

Electric Warehouse Power Marketing came into existence to fill a product need in an already mature market. Power production has reached the mature stage in the product life cycle. Many rivals are now competing for a small portion of the product market. These rivals have basically eliminated new competition because of the huge capital investment that is required for entry and the decreasing margins that this rivalry has caused. Companies are looking for ways to differentiate their products from the competitors by developing niche product strategies so that direct competition is not needed.

THE EWPM BRAND

The branding of the Electric Warehouse product will focus on the needs of the customer and not so much on what actually takes place in providing the power. We want the customer to know that because of what we do they have power when they need it and at the most affordable price possible. We want our name, Electric Warehouse, to evoke a feeling of confidence, trust, and the image of cutting edge technology and customer convenience.

While our competitors continue to produce and provide power in the traditional way, Electric Warehouse is lowering costs to the end consumer by acting as a marketer/broker as well as a producer. Further, consumers will know that when they are served by Electric Warehouse service will be reliable.

Competitors still produce power when the consumer needs it as the consumer needs it. This can cause the use of less efficient equipment at peak times and thus a higher rate. By supplementing our production with wheeled power, Electric Warehouse customers pay less. Electric Warehouse customers will feel that we have their best interest in mind when we develop an enhancement to our services. This cutting edge image both behind the scenes and at the point of contact with the customer will allow Electric Warehouse to develop an image of being an innovator in a market that has been the same for decades. As deregulation moves forward and customers have a choice as to where they get their power, Electric Warehouse will already be the brand leader in customer-focused service in the power supply industry. Our goal is to have customers come to Electric Warehouse to provide them with our service because of our reputation.

PROMOTING ELECTRICITY

The promotional approach warranted for this new strategy to the power marketing market will require a multidimensional approach. The target of the campaign will involve both federal and local governmental considerations as well as those of private corporations and the public-at-large.

Given that we are electrical power brokers as well as producers in this effort, a thorough and long-term campaign will be necessary. Our primary goal will be to assure potential customers that our service will be reliable and cost-efficient. In order to assure minimal governmental interference we see the need to make the Electric Warehouse product so valuable that the users (businesses and the general public) will not stand for having power available from any other source.

THE PROMOTIONAL MIX

Our goal will at the outset be targeted at making the general population *aware* of our presence and our intentions to provide, maintain, and assure reliable electrical power with convenient interaction. To do this our primary devices will be to first advertise and then to obtain publicity so that the public will associate reliable electric service directly and exclusively to Electric Warehouse.

We will be looking for mass market penetration through radio, television, and print ads. Our goal will be to present to the public and business community our assessment of how power is now delivered and contrast that against how much better their service can become using Electric Warehouse.

Our corporate symbol will be the light bulb to illustrate both the simplicity of the product and the convenience it affords. Every visual advertisement will include a view of the light bulb. Television ads will start with an unlighted bulb and end with the lighting of the bulb by Electric Warehouse. We will promote Electric Warehouse as the company that will "bring light to their world". The intent is to be simple, involved with their daily lives, and of personal use. Using colors of green and gold we will intend to emphasize an environmental cognizance.

We will market ourselves as a public service. We will sponsor public events such as charity benefits. Likewise, due to our sensitivity to environmental issues, we will openly support an environmentally advantageous position. Environmental groups are capable of raising unwarranted public concern about power generation practices. We see these groups as a barometer of public opinion that we will target to maintain on a high level of awareness.

On a person-to-person level, at the noon hour in six major cities, a campaign will include the handing out of light bulbs with the Electric Warehouse logo in busy downtown business districts. On a corporate level, we intend to make a sizable public stock offering at a low price in order to create an atmosphere of community ownership. Likewise, if the service is adopted and widely used, there will be more of a tendency to support its propagation and use. A liberal dividend will be considered in order to interest the buyers from the outset. In the next five years when we reach the point of considering further expansion, we should have created a history of profit buy-in from the business community's point of view.

Because we see customer service and reliability becoming strong purchasing influences in the United States, we will make a heavy investment in gaining a strong and beneficially popular position in the market. If successful, it is hoped that expansion into other nationwide regions would also be a feasible consideration.

To this end, the sum of $5,000,000 (or 5% of the annual capital budget) will be budgeted for promotion for each of the first five years. We intend to build a fortress position that will not be soon challenged either by governmental or private industry competitors.

DISTRIBUTION CHANNELS

The main distribution strategy will be to research the existing transmission and distribution systems in Electric Warehouseís target markets and attempt to secure capacity needed to deliver the energy products desired by our customers. Electric Warehouse will need to employ independent operator logistics in utilizing existing transmission systems and distribution networks to provide for adequate electricity delivery to customers.

As the deregulated electricity market develops and matures, Electric Warehouse will need to partner with independent system operators and construct system upgrades desired to fulfill any obligations to clients.

The major changes in the distribution system will be in upgrades in transmission systems. Where local utilities once would build in an area as growth rates determined the need, future need may be satisfied by upgrades to transmission systems and logistically managing electricity flows. People just may call us the FedEx of electricity.

HOW MUCH DOES IT COST?

As our position early in the competitive electricity market warrants, Electric Warehouse will need to be a lower cost provider to large wholesale blocks of customers. This position is forced due to the state of the regulated monopoly of transmission and distribution of electricity.

Our pricing strategy will have to vary from customer-to-customer due to the specific requirements of each. Some consumers, such as a hospitals or retail stores, desire uninterruptable, cost-effective energy. Yet others desire the lowest cost energy. As Mr. Freedman of Dupont stated, in given markets, price will ultimately determine the winners and the losers.

In general, Electric Warehouse will target two major segments which will have clearly different pricing strategies. First, Electric Warehouse will target the uninterruptable regional industrial, commercial, governmental, or retail customers. These users will be considered premium customers where reliability will be paramount but still at a competitive price. Second, Electric Warehouse will target the high-volume, low-margin customer, like the Duponts and U.S. governmental offices, where price is the only factor in determining the supplier. Given these two drastically different strategies, Electric Warehouse should be consistent with the tools needed to win in both market segments. The focus again will be on core competencies built for strategic alliances with producers, customers, other brokers, and utilities (in some cases).

STRATEGY ASSESSMENT

As Electric Warehouse nears the close of the 20th century, we stand well positioned to capitalize on trends emerging in the electrical power market. Technological advances have not changed the nature of electrons, but have vastly altered and expanded their uses. The changes resultant from these advances have consequently altered the previous paradigm of producer-consumer relations.

The rapid proliferation of on-line services and interactive communications systems signal the consumer's willingness to interact with outside information services. The companies emerging as successful today have recognized the value of direct consumer interaction and relationships.

This is the strategy of Electric Warehouse. Low-cost consumer products will always have their place, and low-cost electrical power will, of course, be in demand. By integrating ourselves within the industry, we expect to be able to provide both low cost and high service products. The infrastructure demands and cooperative agreements with other producers and suppliers will test our managerial skills. The complex retail market will similarly test our marketing strategy. The lessons to be learned from observing the broadcast and telecommunications industry must not go unheeded.

Without doubt, we represent a product with an increasing demand. Creating a market domestically is not an immediate concern. Instead, our focus should be, and is, on capturing as large a share of the existing market.

To that end, we expect our strategy to be well founded. Because one electron is the same as another, it will be the perceived quality of the supply and the public positioning of the supplier which will determine who will be the winners. Our commitment to customer satisfaction and convenience plays directly to emerging trends in consumer preferences. Low cost is not, in fact, the primary constituent in purchases of consumer services. Our product strategy, promotional strategy, and distribution are all focused on this determinant.

That we expect to deliver a lower cost product as well also positions us well to promote our service qualities. High quality does have its price sensitivity, and we expect to deliver our product within consumer preferences for cost. A key element of our promotion will be to integrate in the consumer's mind that electricity is not the product by itself. What we are selling and they are consuming is the service as a whole process.

We are confident that our marketing strategy for Electric Warehouse is a good fit with consumer preferences, internal competencies, and our corporate identity. When consumers turn on the lights, we'll be there!

Anderson Consulting, Arthur Anderson & Co., S.C., *Delivery,* Issue 3. 1995, pp. 26-31.

Ban, Stephen D., "Challenges for Gas Technology in a Restructured Electric Industry." *The Electricity Journal,* May 1996, pp. 26-27.

Christensen, Peter C., *Retail Wheeling: A Guide for End-Users.* PennWell Books, PennWell Publishing Co., Tulsa, OK, 1995.

Johnson, Andy, "Information Technology Energizes a More Efficient Bulk Power Marketplace." *Electric Light & Power.* October 1995, pp. 19-20.

Moody's Investors Service, *Electric Industry Outlook."* October 1995, pp. 6-9; pp. 13-14; and pp. 16-17.

National Independent Energy Producers, *The Competitive Power Revolution: Independent Energy's Expanding Role in Electricity Generation.* October 1994.

National Regulatory Research Institute, *Overview of Issues Relating to the Retail Wheeling of Electricity.* May 1994, pp. 72-74.

Pacific Gas & Electric Company, San Diego Gas & Electric Company and Southern California Edison Company, "WEPEX: Independent System Operator and Power Exchange Proposal, Overview." April 29, 1996, pp. 5-9 and pp. 21-25.

Parks Associates, *Electric Utilities: Regulatory, Financial, and Social Forces for Industry Change.* 1995, pp. 4:3 to 4:6.

Prudential Securities Equity Research, *Portfolio Managers' Wrap-Up.* March 8, 1996, p. 3; pp. 26-31.

Rodgers, Paul, "FERC Orders Sweeping Changes for Electric Utility Industry, Requires Wholesale Market to Open to Competition." National Association of Regulatory Utility Commissioners *Bulletin, Number 18.* April 29, 1996, pp. 15-16.

Rudden, Richard J. and Gilani, Rehan, R.J. Rudden Associates Inc., *Effect of Electric Industry Restructuring on the Competitive Price Position of Natural Gas*. February 1996.

Tessmer, Raymond G., Jr., Boyle, John R., Fish John H. III, and Martin, William A., *Cogeneration and Wheeling of Electric Power:* Opportunities In A Changing Market. PennWell Books, PennWell Publishing Co., Tulsa, OK, 1995.

The Arrington Group, *Electric Utility Industry Survey Report: The Response of the Electric Utility Industry to Marketplace Changes*. 1995.

Tussing, Arlon R., and Tippee, Bob, *The Natural Gas Industry, Second Edition*. PennWell Books, PennWell Publishing Co., Tulsa, OK 1995.

Washington International Energy Group, "The 1996 Electric Industry Outlook." 1996.

INDEX

B

C

D

E

F

G

H

I

J

K

L

M

N

Q

R

S

T

U

V

W

Y

It's easy to do the right thing.

CCC makes it simple, efficient, and cost-effective to comply with U.S. copyright law. Through our collective licensing systems, you have lawful access to more than 1.7 million titles from over 9,000 publishers. Whether it's photocopying, electronic use, or the emerging information technologies of tomorrow—CCC makes it easy.

Call 1-800-982-3887 ext. 700 to find out how CCC can help you to Copy Right!℠

Copyright Clearance Center®
Creating Copyright Solutions

222 Rosewood Drive
Danvers, MA 01923

Copyright Clearance Center and the CCC logo are registered trademarks of Copyright Clearance Center, Inc. within the United States